"CAP" CORNISH, INDIANA PILOT
NAVIGATING THE CENTURY OF FLIGHT

"CAP" CORNISH, INDIANA PILOT
NAVIGATING THE CENTURY OF FLIGHT

RUTH ANN INGRAHAM

Purdue University Press
West Lafayette, Indiana

Copyright 2014 by Ruth Ann Ingraham.
Printed in the United States of America.
23 22 21 20 19 18 17 16   2 3 4 5

Library of Congress Cataloging-in-Publication Data
Ingraham, Ruth Ann.
"Cap" Cornish, Indiana pilot : navigating the century of flight / Ruth Ann Ingraham.
     pages cm
Includes bibliographical references and index.
ISBN 978-1-55753-684-6 (pbk. : alk. paper) — ISBN 978-1-61249-337-4 (epdf) — ISBN 978-1-61249-338-1 (epub) 1. Cornish, Clarence, 1898–1995. 2. Air pilots—United States—Biography. 3. Aeronautics—Indiana—History. I. Title.
 TL540.C744I54 2014
 629.13092—dc23
 [B]
                              2013042310

Notice

Every effort has been made to obtain permission and give credit for photographs used in this book. The photographers and the sources of visual material are indicated in the captions. If there are errors or omissions, please contact Purdue University Press so that corrections can be made in any subsequent edition.

I dedicate this book to the memory of my parents, Lois Watterson Cornish and Clarence Frank Cornish. They nurtured me, loved me, and by their example taught me to never stop learning, to revere the natural world, and to be aware of and honor, through photography, the beauty that surrounds us in its multiple forms.

Contents

    Preface, *ix*
    Acknowledgments, *xv*
1. Growing Up Hoosier, *1*
2. His Head in the Clouds, *19*
3. Back to Earth, *41*
4. The Lure of the Skies, *67*
5. Fellowship Forged through Flight, *97*
6. How to Grow an Airport, *115*
7. A New Baer Field and a Struggling Old, *137*
8. Keeping the Home Skies Safe, *149*
9. Calming the Turbulence, *165*
10. Culmination of a Life in Flight, *179*
11. Never Call It Quits, *209*
    Cap's Last Flight, *217*
    Recognitions Earned by "Cap" Cornish, *218*
    Notes, *219*
    Index, *253*

Preface

THIS BOOK TELLS the story of my father, Clarence "Cap" Cornish, a pilot who lived through all but the final five years of the Century of Flight. It also looks at the evolution of aeronautics in Indiana between the end of the First World War and the mid-1950s, the years when Cap was most active in the aviation field. During this period of sputters and spurts, sod runways gave way to asphalt and concrete; navigation evolved from the use of railroad tracks (the "iron compass") for guidance to radar; runways that once had been lit at night with cans of flaming oil shimmered with new multicolored electric lights; and passengers, who once had sat crammed next to mailbags in open-air cockpits, now rode comfortably in streamlined, pressurized airliners.

In April 1917, at the age of eighteen, Cornish volunteered to fight for his country in the Great War and joined the U.S. Cavalry. But it was flying an airplane and not riding a horse that captured his imagination, and within a few months he was in the Aviation Section of the U.S. Signal Corps preparing to be a pilot. In May 1918, while in flight training at Park Field in Millington, Tennessee, he soloed in a Flying Jenny for the first time. Within two months he had earned the wings that he aspired to, which he wore proudly on his uniform. A short time later he was instructing others in aerial gunnery and pursuit at Carlstrom and Dorr Fields in Florida. He had hoped to go abroad to be part of the air battles being waged against the Germans, but he never got that chance: the "war to end all wars" concluded on November 11, 1918, and the following January, Cornish returned to his hometown of Fort Wayne, Indiana.

The glory of being a flyer was behind him, but only temporarily. Three and a half years later, he attended training camp as a reserve in the U.S. Army Air Service. He was back in a Jenny—and back in his element.

Reconnected with his love of flying, Cornish became increasingly involved in aviation-related matters in Fort Wayne. As his stature in the community grew, the local citizens, aviators and non-aviators alike, began to look to him for his expertise and his ideas about local aviation development. He was a strong advocate for a municipal airfield and for the establishment of airmail service through the city. He also was flying actively again. In addition to working as a commercial pilot for various entities over the years, he performed aerobatics, competed in air races, and promoted aviation through his participation in air tours and air cruises throughout the state.

In 1934 Cornish was selected by the Fort Wayne Board of Aviation Commissioners to manage the Paul Baer Municipal Airport. For the next seven years, the facility thrived and expanded under his leadership. When it appeared that the United States would soon be forced to join the fighting in World War II, however, his talents were needed elsewhere. Much had been learned about the use of airpower for military purposes since the First World War, and when Cap was called back to active duty in 1941, he had the opportunity to serve and protect his country as chief of the Flight Operations Division of the Army Air Forces. Working closely with the War Department, the Navy Bureau of Aeronautics, the Defense Commands, the Civil Aeronautics Administration and Civil Aeronautics Board, and even the U.S. Weather Bureau, he helped to defend the homeland by coordinating air traffic and the various uses of airspace during the war years.

When the war ended, Cornish's unique combination of skills and experience made him the natural choice when Governor M. Clifford Townsend was looking for someone to assume the directorship of the newly created Indiana Aeronautics Commission. For the next eight years he worked tirelessly to advance the development of aviation both statewide and nationally. His lifelong dedication never wavered, but in 1952 his tenure was cut short by politics. When a new governor took office and decided to put someone else at the helm of the commission, Cornish found himself out of a job. It spelled the end of his professional career in aviation.

Without a plane at his disposal, Cap's wings were clipped, but in 1968, on the fiftieth anniversary of his first flight, he rented a Cessna 172 to commemorate the occasion. Friends and family gathered at the Indianapolis Metropolitan

Airport, and some of us got to go up for a spin. It became an annual tradition, and as the years rolled on, his following grew from that original small group to a crowd that included TV reporters and newspaper columnists. On December 4, 1995, my ninety-seven-year-old father broke his own record set three and a half months earlier and was acclaimed for several years by *Guinness World Records* as the world's oldest actively flying pilot. He died only three weeks later—and his history was left for me, his only child, to write.

I lived through part of his active career and had heard his stories often, but it was not until the spring of 1996 that I became genuinely drawn into it. That is when I opened the hinged lid on a plain but finely crafted wooden box in my parents' basement. Inside, along with a few personal items, were 125 well-scripted letters in their original envelopes that my father had mailed home to his family in Fort Wayne during his service in World War I. Dated between May 1917 and January 1919, they describe nearly every aspect of those formative years he spent away from home between the ages of nineteen and twenty-one. I soon realized that this bundle of correspondence was a mere fragment of what I had inherited. I moved everything to my own basement, where my Ping-Pong table soon sagged under the weight of boxes full of original documents that my father had saved. This treasure trove of materials included two pilot's books, covering all but three years of his flying career; correspondence that he had impeccably stored in three-ring binders; texts of the speeches that he had given to various groups over the years; photographs that illustrated his life in flight; flying manuals from his years in the military; and hundreds, if not thousands, of news clippings. That material, plus my own research into such topics as events of the early years of flight, the First and Second World Wars, conflicting philosophies about airplanes in wartime, and early aviation luminaries, provided the background I needed to put his contributions to aviation in context. Many of his saved clippings were undated, often with their sources not identified, so I spent weeks at the Indiana State Library, where I scanned microfilms of Fort Wayne's two primary newspapers in an attempt to track down as many of the missing sources as possible—often gleaning new information in the process. *"Cap" Cornish, Indiana Pilot* is the result.

My father did not have advanced degrees; his formal education ended after twelfth grade. Life's experiences were his teachers, and he applied that knowledge to new challenges, even if it meant taking risks. New technology drew him like a magnet, everything from wireless telegraphy to navigational instruments to the latest developments in aircraft science, and even the computers being used by businessmen he knew when he was well into his nineties. He was articulate, resolute, determined, resilient, dependable, and a strict disciplinarian, all personal traits that he applied to the tasks before him—and often expected of others. In the words of a friend, "When he was right, he was right, and he stuck to his guns and his principles." He didn't look back, but he also didn't burn bridges.

His eldest granddaughter, Lisa Krieg, remembered him through a different lens:

> On December 22, 1995, I lost a great Grandpa and others mourned the loss of an elderly gentleman with a twinkle, a knack for story, and a passion for correctness and honesty. He was a great man and a loving Grandpa, someone I could count on; his love was solid. He was frustrating and stubborn and inflexible, but he was charming and funny and sharp. His spirit prevailed to the end of his life.

It was not always easy being Cap Cornish's daughter, but I learned much from his example—to pick myself up after a fall, dust myself off, and go on. I also learned stick-to-itiveness. How else would I have finished this book, a project that took eighteen years from beginning to completion?

As a Purdue graduate, I am grateful that the Purdue University Press offered to help me share this story. My father would be extraordinarily pleased, too. His connections to Purdue extend back at least to the mid to late 1940s, when he attended and gave speeches at seminars offered within the Air Transportation Program of the new School of Aeronautics. In 1952 the Aeronautics Commission of Indiana, which he then directed, announced plans to meet with officials at both Purdue and Indiana University to discuss the possibility of establishing an aviation library and museum for the state, envisioned as a "fitting memorial" to Hoosiers who had made outstanding contributions to aviation.[*] Although nothing came of the proposal at the time, it was an idea that he continued to pursue. In 1990 he and others, notably his close friend and fellow pilot Edmund F. Ball, suggested again that Purdue provide institutional space to house the histories and biographies of early Indiana pilots. They exchanged letters with Dr. Alten F. Grandt Jr. of the School of Aeronautics and Astronautics and also met with him in West Lafayette. Everyone involved agreed that Purdue University would be the ideal place to preserve and showcase Indiana's

---

[*] "Aviation Museum Proposed for State," *Indianapolis News,* November 19, 1952, 25.

rich aerospace heritage. The Purdue Libraries are now home to the Barron Hilton Flight and Space Exploration Archives. It is with great pride that through this book and my partnership with Purdue I can contribute a chapter to the literary portion of the history they envisioned.

Ruth Ann Ingraham
August 15, 2013

Acknowledgments

My father and mother meticulously saved and protected documents and artifacts that span a century in time. When I inherited their voluminous collection in 1996, I became the keeper of the ingredients for a story that had not been told, and I committed myself to carry it to fruition.

Making sense of all their materials was more than I could manage at first, so in the summer of 1997 I asked Mark Copple, a neighbor who had just graduated from college, to begin the organizational process. He put dozens upon dozens of undated newspaper clippings in chronological order—a monumental task—and assembled them along with other materials. He also taped interviews with pilots, friends, and relatives—people who knew my father and who related their memories of him. Mark's work was a vital first step and a huge help.

More than a decade later, when I began to pull the elements of my father's story together, I turned to Nancy Niblack Baxter, Indiana historian and author, who advised me about how to write a biography and who was my cheerleader.

Along the way, I asked friends to search for errors and omissions. Ed Elrod answered questions about wireless telegraphy and early radio, an interest to which my father devoted significant time when he was in his teens and twenties. Richard Cunningham read the chapter about the Aeronautics Commission of Indiana, for which he had been director in the 1960s. Roger Myers, author and aviation museum curator, reviewed the chapters about the airfields in Fort Wayne in the 1920s, 1930s, and early 1940s. Steve Baranyk checked references to World War II. My long-lost step-cousin David G. Ehrman, a Fort Wayne pilot who is intimately familiar with both Smith and Baer Fields, read a later stage of the book in its entirety. Don Manley, pilot and airport engineer, answered more questions for me as I neared completion of this project and helped me select illustrations to accompany the text. John Schalliol reviewed the text from start to finish and gave the book's factual information his "stamp of approval." My cousin William A. Welsheimer, who frequently rode out to Paul Baer Municipal Airport with my father, his uncle, has often shared with me his vivid recollections of those times. For instance, he rode with the men who hung and lit the kerosene lanterns at dusk to mark the landing field for nighttime flyers. I thank them all for sharing their time and knowledge.

Jane Lyle is my book editor for the second time. We share the same goals—that my book be a pleasure to read and also informative and accurate. She offered me solace and encouragement when I felt overwhelmed. She scrutinized every word I wrote and challenged me to examine, clarify, and refine. I could not have completed *"Cap" Cornish, Indiana Pilot* without Jane, a brilliant woman and friend.

Behind the scenes were my daughters, Christy Krieg and Lisa Krieg, who gave me their love, their unwavering enthusiasm, and their belief in this project. They were a beloved part of their grandparents' lives, and are of mine.

Finally, I thank Charles Watkinson, director of Purdue University Press, and his exuberant staff for allowing this literary project to take flight.

* * *

As is true for many of us, our working years can be consumed by our jobs. That was true for at least the first decade that followed my parents' marriage in 1926. My father loved all things pertaining to flight, and he totally immersed himself in many aspects of it. The fact that during the months of July and August 1929 he flew every single day—sixty-two days straight—is just one example. In the 1950s a job change shifted the tenor of their lives significantly, and by the 1960s they were able to enjoy shared interests—their home and garden, photography, travel—and could spend time watching their two adored granddaughters grow to adulthood. They remained both curious and creative through most of their long lives, and I am grateful for their love and for the multitude of qualities that helped shape who I am.

1

Growing Up Hoosier

THE DAY BEFORE Clarence Cornish's tenth birthday in 1908, he had just pitched some hay to Brown, the family's horse, when he tumbled from the haymow and landed on the barn floor. The tine of the pitchfork he was using shallowly pierced his face a fraction of an inch from his eye. Fortunately, chance left him with only a flesh wound. His vision spared, he would go on to become a pioneering pilot who helped shape both Indiana's and the nation's early aviation history.

On a sunny May morning in 1918, the wings of a flimsy Flying Jenny aircraft, aided by an OX-5 engine, lifted Cornish from the earth for his first solo experience. Seventy-seven years later, following activities in aviation that spanned most of the Century of Flight, he would pilot a plane eighteen days before his death at the age of ninety-seven. That feat would bring his career to a glorious conclusion, earning him posthumous recognition in the *Guinness Book of World Records* as the world's oldest actively flying pilot.

Clarence Franklin Cornish, the only son of Frank and Ada Buck Cornish, was born November 10, 1898, in St. Marys, Ontario, Canada. His life as a Canadian was brief. On October 13, 1900, with his parents and older sister, the towheaded toddler boarded a passenger train on the Grand Trunk Railway, headed for a new life in the United States. After stops at some small towns west of St. Marys, the Cornishes crossed the border at Port Huron, Michigan. They then took a second train, on the Lake Shore and Michigan Southern Railroad, to Detroit, and finally a third, on the Wabash Railroad, to Fort Wayne, Indiana.[1] It was there that Cornish would one day help direct the city's aviation evolution.

Clarence's paternal great-grandfather, William Cornish, had emigrated from Stratton, Cornwall, England, in the 1830s and settled near the town of Bowmanville, Ontario, located east of Toronto on the northern shore of Lake Ontario. In 1838 William married Eliza Frank; on a one-hundred-acre farm northeast of town, they grew wheat and raised ten children. The first of those children, John, eventually left the area and moved to the western part of the province. There he married Jane White, who had been born in 1844 in County Tyrone, Ireland, and had immigrated to Kirkton, Ontario, with her family when she was nine years old. John and Jane produced seven children of their own, two of whom died in infancy and one at age five. The birth certificate shows that the fourth of those children, Franklin, was born in 1870 in nearby Mitchell, a community about ten miles from the family's farm.

In his teen years, Frank chose to go into the painting and decorating business (even embellishing the spokes of carriage wheels) instead of following the farming tradition of his father and grandfather. He practiced his trade east of the tiny settlement of Kirkton in the larger town of St. Marys, presently nicknamed Stonetown for its limestone buildings that date to the latter half of the nineteenth century. Frank remained a professional painter throughout his long life.

Ada Buck, Clarence's mother, was born in Indianapolis in 1869, one of the six children of Charles W. and Louisa Durfee Buck. Ada's paternal grandfather, William Buck (who had changed his last name from Clutterbuck), born in 1800, had emigrated from England to

> Ada boarded as a non-Catholic student at the Academy of Our Lady of the Sacred Heart. According to an advertisement in *Kramer's Business Directory*, the school for "persons desiring to acquire a solid, useful and ornamental education" was "a twenty minutes' ride from the city, on the Ft. Wayne & Jackson Railroad." Subjects ranged from astronomy to "Higher Mathematics" to "Belles-Lettres." Boarding students were expected to bring with them "4 towels; 4 napkins; 1 knife, fork, spoons, goblet; 1 dressing gown; 2 black aprons; 4 changes of linen; summer and winter balmoral [a type of wool petticoat]." The cost of board and tuition for a five-month session was $75, with additional fees for laundry and bedding, for classes in music (harp, guitar, and piano), modern languages, Latin, drawing, and painting, and for "Artificial Flower Les'ns" and "Artificial Fruit and Leather Work."[2] A good student, Ada received the book *Catholic Flowers from Protestant Gardens* for her outstanding performance in many areas, including "Epistolary Correspondence."

Saratoga, New York. In adulthood, William's son Charles became a railroader; he worked with the Peru & Chicago Railway Company as a fireman from 1860 to 1867. Louisa, born in 1832 in Sylvania, Ohio, married this Charles in 1855, and together they raised six children. After her death in 1880, Charles moved his children from Indianapolis to Fort Wayne, where Ada attended the nearby Academy of Our Lady of the Sacred Heart boarding school.

In her early twenties, Ada lived with her eldest sister, Lucy Buck Crabbs, in Lapeer, Michigan, where it is assumed that she first met Frank Cornish, who may have been there to fulfill a painting contract. After a period of courting, Ada accepted Frank's marriage proposal and sent her father a letter with the news. Charles answered with these words:

> Fort Wayne Apl 23 - 1892
>
> My Dear Daughter,
>
> Your welcome letter at hand and glad to learn that you had regained your health. Ada you have indeed given me a surprise knowing the futility of giving advice in such cases I will refrain from so doing only to say that if he [Frank Cornish] is your choice that I wish you unbounded happiness. You do not say when or where the happy event takes place but of that I shall expect to hear more hereafter. I enclose you herewith a draft for $25.00 which I trust will do you for the present.
>
> Yours as ever
>
> Your Father[3]

Charles supported Ada's decision to marry, but reluctantly, it would seem.

On December 18, 1892, Frank Cornish and Ada Buck were married in Fort Wayne in the Buck family home. The newlyweds lived in St. Marys, where their first child, Irma Katherine, was born in 1894. Fortunately for the young family, Frank Cornish's clientele grew, and he ran the following announcement in the January 23, 1896, edition of the St. Marys *Argus*:

> Change of Business
>
> I wish to announce to the public of St. Marys and surrounding country that I have purchased the painting, paper hanging and decoration business

Ada Buck and Frank Cornish married on December 18, 1892, in Fort Wayne, Indiana. (Author's collection)

from Mr. Gilbert White, and am prepared to do everything in my line in first-class style. I have over 10 years' experience and understand it thoroughly, so that all favoring me with work may depend on getting satisfaction. House painting, paper hanging and decorating attended to promptly. Shop at the old stand—Queen Street, West Ward.

FRANK CORNISH

Ada bore a second daughter, who died in infancy in 1897. Clarence, who would turn out to be his mother's favored child, came along a year later. For the next three years, the family continued to live in the attractive Ontario community situated on the banks of Trout Creek and the Thames River not far from Stratford and Shakespeare. On a late autumn day in 1900, however, they left Canada for reasons unknown. Perhaps Ada found it difficult to be separated from her family by such a great distance. Perhaps she missed the more sophisticated bustle of the Indiana cities in which she had grown up. Perhaps Frank found it difficult to support his growing family there, although St. Marys was experiencing a commercial and public building boom at the turn to the twentieth century.[4] Fort Wayne may have seemed like a better place to build a clientele.

The Cornish family boarded a train at this station in St. Marys, Ontario, Canada, and settled in Fort Wayne, Indiana. (Author's collection)

Whatever the family's motivations, Frank, Ada, Irma, and Clarence emigrated from Canada for a future in America.

At first the Cornish family boarded with Buck family members on Franklin Street ("way out in the northwest section of town," as Clarence once described it), and Frank worked in Bucks' Baltimore Fish and Oyster House. But he soon started another painting and decorating business, and he and Ada were able to establish their own household on Runyon Avenue in a middle-class neighborhood on Fort Wayne's east side. Frank's commitment to excellence and to detail quickly gained him appreciative customers. He became known as a skilled craftsman, particularly for his work in intricate stenciling applied along the upper edges of interior walls, a fashionable décor element at the time. For a more opulent accent, he and other artisans added ornamental magic, including gold leaf highlights, to the Bass mansion, Brookside, which was then on the outskirts of Fort Wayne. Built for the family of industrialist John Henry Bass in 1903 and referred to by some as "the Castle," the former residence is now part of the University of Saint Francis campus. When Frank was not earning a living for his family, he was, as Clarence later remembered, quite a photographer as well.

Back within the embrace of her extended family, Ada nurtured Irma, Clarence, and then a third child, Marcia Odetta, who was born in 1906 at the

Cornish family friends join Clarence (left) and his sisters in a hammock around 1906. His older sister, Irma, is holding their baby sister, Marcia. (Author's collection)

family's next home at the northeast corner of Main and Fry Streets. Pregnancy was a taboo subject in that post-Victorian era, so seven-year-old Clarence was unaware of the impending birth until a February day when he got up one morning and learned that there was a newborn sister in his parents' bedroom. Two years later, the Cornishes moved again, this time to Hoagland Avenue. Frank partnered with John W. Bowers to form a new business, Cornish & Bowers, Painters & Decorators, located initially on Lincoln Avenue and later on Broadway. The family moved in 1914 to Packard Avenue (where they had indoor plumbing for the first time), and finally, in 1916, to Thompson Avenue.

Life was not always idyllic for the Cornish children. Frank was a strict disciplinarian, and he did not dote on his children the way that Ada did. "There was very little conversation or communication," youngest daughter Marcia said later in life. "We were simply five people living under one roof. . . . My father didn't talk to me or show any affection."[5] Irma went to work when she was only sixteen, contributing to the family's income by working as a clerk, typist, and stenographer for small companies such as Kidd & Company, the Letter Shop, and the Wildwood Builders Company. She subsequently joined the New York, Chicago and St. Louis Railroad, where she added accounting to her tasks. Despite the sometimes strained dynamics in the Cornish household, the siblings certainly had their lighter moments. For example, family members recall Clarence describing a Halloween prank that involved greasing trolley tracks with soap so that the car would slide past its intended stop.

Photographs of Ada as an adult show an angular, tight-lipped, stern-faced woman, very different from the sensual, romantic girl that her early writings reflect. In one photo, she and Frank stand well separated, rigidly detached, without a glimmer of affection. Perhaps something had occurred during their lives together that led her to write:

> As I get older and years come and go, my mind wanders back to those other days, to the many things which have happened, and I wonder, if there was a land of beginning again and the same things happened, would I conduct myself in the same manner, under the same circumstances. I wish in this land of beginning again I could throw off the unpleasant things which have happened and, like an old garment, be thrown away and never put on again.[6]

Did socioeconomic differences—Ada's affluent, educated, more urban background versus Frank's agrarian, small-town upbringing—lead to misunderstandings, slights, and disappointments? When Ada's writings are compared with letters that Frank wrote late in his life, it is clear they had vastly different verbal skills. Nonetheless, the two were married for forty-four years.

Like other children of modest means in the early 1900s, Clarence was expected to do domestic chores along with his schoolwork. But there was also time for play and socializing. He was fascinated with new technology and enjoyed experimenting and tinkering. And like many active young boys, he experienced his share of mishaps. A year after the fall from the hayloft that gashed his face, he took a misplaced step onto a rusty nail that led to an alarming infection and painful treatment, requiring a dose of chloroform while the family doctor cleaned the wound.

In 1908, the doctor who made house calls to care for the boy's injuries would most likely have arrived in a horse-drawn buggy. But horse power was already starting to give way to horsepower, as new technology was ushering in the automotive age. A mere four years earlier, the January 1904 issue of *Frank Leslie's Popular Monthly* had probed the efficacy of the automobile and predicted that this mode of transportation would one day be used by "the poor man . . . as surely as the rich man." The article illustrated eighty-eight gas-, steam-, and electric-powered models on the market in the United States for those who could afford one.⁷ Undoubtedly the number of manufacturers and models had increased significantly by 1908, and it would not be too many years before the average family's horse in the barn was replaced by a car in the garage.

Powered flight had also become a reality, as most famously represented by Orville and Wilbur Wright's success with their *Wright Flyer*. But airplanes were still in their infancy, and no one could have imagined at the time how they would eventually transform the world. Later in life, when the Wright brothers were asked what had originally sparked their interest in flight, they pointed back to a simple but fortuitous moment in their childhood: in 1878, while the Wrights were living for a time in Iowa, their father, Milton, brought home a plaything known in French as a *hélicoptère*, designed by Alphonse Pénaud.⁸ When the rubber bands connected to paper-covered rotors at each end of a bamboo stick were twisted and then released, the rotors spun in opposite directions and propelled the flying device into the air. The brothers never forgot this toy, which in one way led to their triumph with the *Flyer* twenty-five years later.

When Cornish was asked what had first triggered his interest in flight, he related a similar experience. He was around eight years old, he said, when, shortly after the Wright brothers' first powered flight, "a company started manufacturing airplane parts which could be assembled into a model airplane. How I managed to purchase the parts for a Wright airplane powered by rubber bands driving two propellers I do not know. I assembled the parts as described but it would not fly."⁹ The energy from the released propellers was too rapid, and though the propellers turned "like heck," the plane went nowhere because "it was just one [quick] shot." Since it would not even glide, he gave up on the whole idea. Although there is no indication that Cornish ever mentioned it, he may also have been influenced by the excitement surrounding a group of aviators who appeared in Fort Wayne in 1910, when he was eleven. They were part of a tour being sponsored by aviation pioneer Glenn Curtiss's exhibition company. Among the flyers that day was Blanche Stuart Scott.¹⁰ Not only was she the first woman pilot to perform in an air show, but she was making her debut appearance. And Cornish must have known about local hero Art Smith, the "Bird Boy," who made headlines in 1911 when he flew his plane to nearby New Haven. "There was a big crowd out that day," Smith wrote in his autobiography. "People in Fort Wayne had been growing more interested in my attempt to fly.

Several of the members of the original Fort Wayne Boy Scout troop, 1910.

Clarence and other members of his American Boy Scouts chapter proudly wear their new uniforms around 1910. Cornish is in the second row, second from the left. ("Several of the Members of the Original Fort Wayne Boy Scout Troop, 1910," *News-Sentinel*, Rotogravure Supplement, February 10, 1934. Used with permission from the [Fort Wayne, Ind.] *News-Sentinel*.)

I suppose a great many of them expected to see a bad accident. The newspapers had reporters there."[11]

When Cornish was young, however, flying was not his main interest. Between the ages of twelve and eighteen, his primary extracurricular activities involved the American Boy Scouts (ABS) and amateur radio, both of which strongly influenced his adult life. He spoke often and passionately about his years in the ABS, a fledgling organization that emphasized diligence, dedication, teamwork, patriotism, and learning to rough it outdoors.[12] A few prominent Fort Wayne citizens, among them Edwin T. Jackson, formed the local chapter around 1910, and Cornish enthusiastically joined a few months before his twelfth birthday.

Uniforms were required for membership in the ABS, so the members of the new chapter, who included a future brigadier general, Robert Porter, began working to raise the needed money. As one means toward that end, they collected newspapers—so many that they "filled a downtown building with them." Their mission accomplished, the group members stood proudly with their leaders on the steps of the Fort Wayne Central Library, posing for a photo in their new outfits. "They fit those uniforms like a shawl on a pump, but we were very proud of them," Cornish later recalled. In 1913 they learned that the American Boy Scouts were now to be known as the Boy Scouts of America, and they would be required to purchase a different uniform. "We said to heck with that. We're going to wear these uniforms whether they like it or whether they don't." This seemingly minor affront left a deep impression on Cornish, one that he still felt eighty years later, a reflection of his deep sense of fair play and his willingness to defend his convictions in the face of obstacles. Those traits would serve him well in his adult endeavors as a pilot and aviation leader.[13]

The patrol, also referred to by Cornish as "the group," met weekly on South Wayne Avenue west of Harrison Street and practiced "class-order drill," led by a naval petty officer on recruiting duty in Fort Wayne. Their basketball team played in the basement of the Jefferson Street School. At other times the bicycle patrol rode around the countryside: "I rode to Rome City and back," Cornish recalled, "about fifty, sixty miles in a day. Most would not ride that far with me." Through scouting, future leaders developed self-discipline, a spirit of adventure, and the ability to meet individual challenges.[14]

Scouting also encouraged stamina and resourcefulness, which the boys put to use on camping expeditions. Robison Park had opened in 1896 along the St. Joseph River above a feeder dam. It was an instant success, and people from the city rode an open-air trolley the seven-mile distance into the country to enjoy a myriad of summer pleasures—boating, amusement rides, dancing, picnics, and entertainment. The members of Cornish's Scout troop were permitted to set up their tents in an undeveloped section of the park, where they literally learned the ropes. Cornish recalled an incident when everything was still new to them:

> A big storm came up. So we tightened the tie-down stakes, preparing for the storm. Well, the storm hit and down went our tents. Why? Because those ropes shrunk and they just pulled the stakes right out of the ground. The tents went "bleh." Our mess tent was a big heap. After the storm was over, we went back and did it all correctly. And we learned that you don't tighten up those doggone ropes. They were not nylon and they were subject to shrinking when they got wet.[15]

A year or two later, the Scouts moved camp from Robison Park to a forested site along the shore of Sylvan Lake at Rome City. Cornish recalled camping

Clarence, far left, and other members of his BSA troop attend camp at Sylvan Lake, Rome City, around 1912. (Author's collection)

activities in the woods with some of the older boys who were more experienced. "We had a marvelous time there," he said. This included swimming.[16]

An average student at Fort Wayne High School, Cornish was not involved in any activities that set him apart from most others in his class. Still, he was a busy and productive young man. For a nickel each, he walked miles to deliver the *Saturday Evening Post* and the *Ladies' Home Journal* to his grandfather's house, which was "way out" at 2405 Calhoun Street. For two summers during his high school years, he worked at the Perfection Biscuit Company. Now Perfection Bakeries, the company was founded in 1901 and was the first to offer wrapped and, later, sliced bread to customers in the area. He worked "a good full six days a week for $3.00." The second year he received a 50 cent raise, a good income by his standards. To pay school expenses during his junior and senior years, he delivered a morning paper. After getting up at 3:00 a.m., he rode his bicycle uptown to the *Journal-Gazette* newspaper office, where he picked up the papers and headed out on his route, pedaling "out in the northwest corner of town, into the countryside to the old Bass farm, then back into town. Then clear out southwest to my house and then ride to school. In class I was asleep half the time or sending notes to one of my buddies."[17]

Cornish was almost sixteen when amateur radio first grabbed his attention and his time. From the family's Packard Avenue home, he began transmitting signals

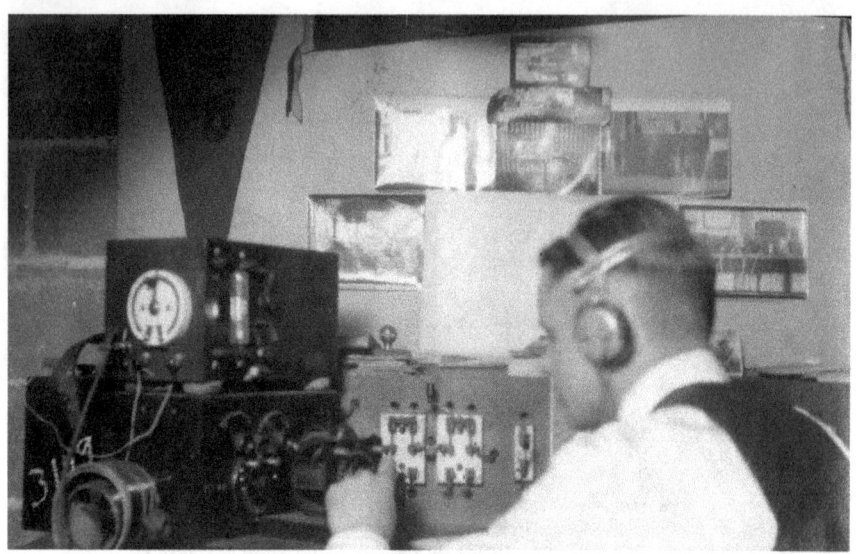

Clarence communicates by wireless telegraphy from his home in 1915. (Author's collection)

"using a Ford automobile spark coil and a crystal detector for receiving radio transmissions." He "later developed a more sophisticated station using a radio tube detector set, which [he] built, and a three-quarter-kilowatt rotary spark gap transmitter with amateur radio station license call 9KG 1914 to April 1917."[18] His innate intelligence and curiosity drove him to experiment with radio in this rough, developmental phase.[19] It was a thrill to connect with people in far-off places. He had no way of knowing at the time that this early experimentation would lead to a full-time occupation in the radio business a decade later. He may have used some of the earnings from his paper route to improve his station equipment when they moved to Thompson Avenue. An article on wireless telegraph stations in the Fort Wayne *Journal-Gazette* dated February 13, 1916, mentions Cornish's station, CAR, reporting that it "has been heard by stations at Indianapolis and regular communication had been transmitted for a distance of sixty miles. Clarence has heard messages from the Atlantic, Cuba and the Great Lake regions."[20]

There were some setbacks during those years. Cornish related this story about an incident that alarmed his parents:

> When we lived on Packard Avenue I recall I had a big flattop antenna out in the backyard that ran from the house out to the barn, and this big flattop antenna was just parallel over the wires that brought the power

into the house. The folks had just bought a brand new solid oak dining room set, and it had an Irish linen cloth on the table. And in the dining room right over the table was a chandelier with several lights in it. I had just gone to a three-quarter-kilowatt transformer for transmitting and I hit the key with that thing and I heard some screams downstairs. The power from the antenna had been picked up by the power line that went into the house and it melted the brass in the sockets in the chandelier, which dropped down, burned holes through the linen cloth and into the top of the brand new dining room table. And if you think I wasn't in the doghouse. We had to do something about that and fast.

His forgiving parents allowed him to move his station equipment out of the garage and into the furnace room after he demonstrated that he could pick up music transmitted by Pittsburgh area station 8XK, later KDKA. "We could actually hear music signals using earphones, though it would fade badly at times," he said.[21]

Cornish was not alone in his passion for radiotelegraphy. He was one of a dozen amateur wireless telegraphy enthusiasts who formed the Fort Wayne Radio Association (FWRA) on July 1, 1915. This organization was made up of the pioneers in the study of wireless in Fort Wayne. In addition to the officers—George Carter, president; Cornish, vice president; E. H. Erickson, treasurer; Morton B. Williams, secretary; and L. C. Young of Van Wert, Ohio, chief operator—they included George Bauer, Louis Hermann, Ross Parnin, DeWitt May, Davis Diffendorfer, and Professor Roscoe Coats.[22] The members, most of whom were students at Fort Wayne High School, were either affiliated with or had applied for membership in the Central Radio Association of Amateurs, "a national organization" for radio buffs "between the Rocky mountains and the Ohio river," according to an article in one of the local newspapers.[23] The article said that there had been three wireless stations in Fort Wayne in April, and now, three months later, there were between twenty and twenty-five. The popularity of this experimental stage in the development of communications was rapidly expanding the medium.

The idea of conversing with friends without the use of an electric wire (consider the cell phone of the twenty-first century) intrigued the FWRA members. Roscoe J. Coats, who had one of the best-equipped amateur stations in that part of the state, and Carter, who taught physics at Fort Wayne High, were instrumental in fostering knowledge about wireless telegraphy among the young men. At semimonthly meetings, they gave demonstrations and lectures about the progress of wireless and led discussions. Carter had successfully conversed with stations on the Gulf of Mexico and "the western mountain range" (presumably the Rockies) and had picked up messages from the Atlantic Ocean. The *Journal-Gazette* credited Carter and Young, an operator for the Pennsylvania Railroad Company, with having brought the work to such an advanced level. Amateurs in the city were hearing messages from all over the United States and even, in one or two instances, from Germany. "This has been the culmination of their efforts and has been recognized by amateurs throughout the middle west as one of the most remarkable incidents in amateur wireless telegraphy in this part of the country." With a degree of local horn-tooting, the *Journal-Gazette* added:

> As with the aeroplane the work has just begun.... It has been by the work of just such amateurs as are daily striving here to perfect their stations that the zenith of all sciences has been reached.... Fort Wayne citizens should at every opportunity encourage their work and lend a helping hand when such is possible. That day has passed when such enthusiasts as these young men are looked upon as theorists who are working with no ultimate end in view. We are beginning to take strides towards that day when any man willing to sacrifice his time and money in such causes will be looked upon as among the most public spirited of the country.[24]

The members of the FWRA had reason to celebrate. They recognized how fortunate they were to be part of a rapidly growing body of people who were developing the capability to communicate for the first time across the miles, and even across national borders. They held a celebratory banquet on January 15, 1916, at the Wayne Hotel, which was followed by a theater party at the Palace.[25] The following spring, the 1917 *Caldron*, the yearbook of Fort Wayne High and Manual Training School, devoted three pages to the FWRA. Even though it was not an official school organization, it was composed of students and recent graduates. Member DeWitt May, in writing about the FWRA, credited his "beloved professor" Mr. G. Carter, physics instructor, as "the real booster of the Club and its progress." He went on to say that the association's aims were to study wireless telegraphy and to find a way to minimize interference (jamming), the greatest problem of every radio organization in the United States in those years. With the ever-increasing number of stations around the city of Fort Wayne, it was necessary to organize and regulate the time of talking to allow everyone a fair chance. Interferers were often referred to as "robbers"

Fort Wayne Radio Association members hold their first annual dinner meeting in 1916. Clarence is second from the right. (Author's collection)

and "thieves" because they deprived other radio users of valuable time and the pleasure of being able to converse with fellow enthusiasts. "Contrary to our rather hateful thoughts and wishes," May added, "we became sympathetic and decided to let them [interferers] in the Club, with the purpose in mind of teaching them by kindly working for them and instructing them instead of by the 'whip' method." The yearbook article concludes with a farewell tribute, a radiogram or series of dots and dashes, to Carter, who was moving to Detroit to take a teaching position there.

> Frustration caused by "interference" was widespread and not limited to Fort Wayne's amateur wireless enthusiasts. The February 1917 issue of *QST*, a popular magazine for amateur radio enthusiasts, included a letter about a "pest" or "ham" who on a "Test Message night" when others "were trying to go north on their Trunk Line" wanted to work long distance at the same time.[26] "He started sending a story book and then said, after half an hour, 'My hand is so crampy now that I will have to quit using it and sit on the key,' which he did for about another half hour. By the time he had finished his 'QRM' it was too late to try anything else on relaying, so we had to quit."[27] The editorial staff of *QST* suggested that something needed to be done to control the absolute stagnation of traffic on account of this nightmare known as QRM, and that a law might be introduced in Washington. "The big noise which bothers us if we sit up late enough is the fellow with the impure wave and the big decrement (the ratio of the amplitudes in successive cycles of a damped oscillation)."[28]

FWRA members Cornish, May, E. Erickson, Ross Parnin, and George Bauer received an invitation from the local Electro-Technic Club to attend the club's meeting on April 20, 1917, and perform a "stunt" for several hundred members gathered on the third floor of Building No. 26, the first of the large new structures being completed by General Electric.[29]

> The success of this stunt rests with the application of a newly invented contrivance called the Multi-Audio-Fone [an amplifier], which so intensifies the electric waves that everyone present may audibly hear them. By the erection of a wireless aerial from the top of the high water tank adjacent to Building No. 26, the messages sent out that night as government relay tests are to be picked up and transferred to the ears of all the Electro-Techs assembled in the building by means of the loud-speaking horn. The relay messages, dispatched only on Tuesday and Friday evenings, under government auspices, are sent from the powerful station at the University of Michigan.[30]

Unfortunately, the stunt never took place. All such activities were interrupted when the United States entered the Great War, at which time the government ordered all amateur and commercial radio telegraphers to shut down their stations. The airwaves would be needed for the war effort. Cornish received a letter from the Department of Commerce, Navigation Service, Office of Radio Inspector, dated April 7, 1917, in which he was ordered to dismantle his equipment. After the war, he resumed operation with station license call 9FB.

The war had begun in Europe in 1914 with the assassination of Austria's Archduke Francis Ferdinand. Preexisting alliances drew countries from around the globe into the ensuing conflict, but the United States remained neutral at first. In 1915, however, a German U-boat destroyed the British ship *Lusitania*, resulting in the loss of 123 American lives. Germany then attacked an increasing number of American cargo and passenger ships, sabotaged factories in the United States, and pledged financial support to Mexico if it would fight to regain "lost territory in New Mexico, Texas, and Arizona."[31] On April 2, 1917, President Woodrow Wilson asked Congress for a declaration of war. Four days later, that declaration was passed.

The residents of Fort Wayne responded with patriotic zeal. Tens of thousands of citizens, a throng "of monster proportions," lined the streets on April 19 for a Lexington Day parade, held to commemorate the opening battle of the Revolutionary War in 1775. That same day, known nationwide as "Wake Up America Day," a national call to arms was issued. Following the parade, there were speeches by civic leaders, and resolutions supporting the war effort were unanimously adopted. At the end of the ceremonies, "Capt. George L. Byroade, U. S. A., raised a large flag to the top of a ninety-foot flagstaff. Then the vast crowd, with heads bared, united in singing 'The Star-Spangled Banner.'" That rousing event marked the beginning of an outpouring of support, and the city's residents answered the call. The local Army recruiting office was especially busy. According to the *Journal-Gazette*, the federal government gave "national publicity to the fact that in the matter of enlistments between the Boston Harbor and the Golden Gate and from the Great Lakes to the Gulf, the Fort Wayne district leads the procession." By the end of April, the local district had already filled a remarkable 75 percent of its quota.[32]

To avert a food shortage and a prohibitive rise in prices, Boy Scouts distributed pamphlets to each home, advising householders that "every man, woman and child who can wield a hoe and make the earth produce food owes it to himself and to all of us and the government besides to do what can be done to increase the crop of food. Remember, we shall pay dearly in the winter for indifference now. America has never known what it is to suffer for food, but America will know to her sorrow unless she plants and sows. Be a patriot—not a 'slacker.'" By early May, Fort Wayne had 3,734 known vegetable gardens, 2,790 of which were located in backyards. A local committee distributed twelve hundred bushels of seed potatoes and thirty bushels of beans at savings, with "a quantity provided free to families in poor circumstances." During the summer months, home economists demonstrated the recommended processes for canning fruits and vegetables. The city even adopted a "daylight saving" plan, effective Saturday, May 12, 1917, to accommodate working gardeners.[33]

Fort Wayne began documenting all unnaturalized German residents over the age of fourteen, and by mid-June, 850 "alien enemies" had been identified, many of whom were prominent in the city's commercial and professional life. They were allowed "to continue to reside within, to come within or pass through, be employed within, or conduct business within, the areas (or zones) within one-half mile radius" of the manufacturing plants of the General Electric Company, S. F. Bowser and Company, and the Bass Foundry and Machine Company, as well as the Pennsylvania Railroad shops and the east car shops, all of which were engaged in government work. Each "alien enemy" was required to carry a permit that bore the holder's photograph.[34]

On April 27, 1917, Cornish and four other young men from the FWRA heard "the call of the nation to follow the Stars and Stripes in the World's greatest war in defense of liberty and democracy" and "enrolled as defenders of the land of Uncle Sam." "Patriotic Members of Radio Club Respond to Nation's Battle Call," proclaimed a newspaper headline.[35] Cornish had taken the first step on a path that would lead him into the air, and eventually to a rewarding life in aviation.

2

His Head in the Clouds

IN HIS PATRIOTIC desire to help "make the world safe for Democracy," Cornish headed first to the Navy's recruitment office in Fort Wayne. Perhaps it was paddling Indiana's northern lakes in a canoe, occasionally slipping through clear waters with a sail attached, that had led him to visualize himself on a ship. Despite having stuffed himself with bananas, however, he failed to meet the weight requirement and was rejected. He turned next to the Army, where he envisioned himself face-to-face with the enemy on European soil. An oft-told story was that Frank Cornish had given his son this sensible advice: "Look, if you've got to join this war, why don't you join the cavalry and ride a horse instead of walking?" Clarence took that advice, joining the 2nd U.S. Cavalry with three of his friends from the Fort Wayne Radio Association—but he never sat in a saddle. Nor were the four friends able to stay together, as they had been promised. "We were together for two weeks," he recalled. When he returned to attend commencement exercises held on June 21 and to receive his diploma from Fort Wayne High School, Private Cornish was wearing a soldier's uniform, ready for combat in the Great War.

Cornish posted a letter to his mother dated May 3, 1917, one of 125 letters that he wrote to her during his twenty-one-month stint.[1] His words that day may have allayed his parents' initial worries. In fact, he was not having the awful time they had anticipated when he went off to war. He was sleeping well, had received his summer uniform, and was finding the food better than expected. "We have had pie 3 times. Of course cookies, cake or candy [from home] will

Soldiers line up outside the mess hall at Fort Thomas, Kentucky, in 1917. (Author's collection)

never be refused." Although military camp may have been reminiscent of happy times in Scout camp a few years earlier, Cornish wrote fondly about his Fort Wayne buddies, a comfort for the eighteen-year-old in new surroundings who had seldom been away from home overnight.

His first assignment was to measure recruits for their uniforms. For this, the military paid him a private's wage of $30 a month. Fitting droves of recruits for their shirts and trousers and doing occasional guard duty may have been monotonous while at Fort Thomas, but Cornish's recreational and social life during his eight months in Kentucky were not. He dove from a twenty-six-foot-high platform into a pool at Chester Park and swam with friends in the Ohio River, "the hardest part" being the climb down and back up the steep bank that bordered the water's edge, which he said took at least twenty minutes. He wrote that while swimming they benefited from waves created by the *Island Queen*, a popular riverboat. On the Fourth of July he danced at Cincinnati's Coney Island, and he continued dancing on the boat that ferried them from Ohio back to Kentucky. He made new friends whose families treated him to meals, including a Thanksgiving feast after which he smoked his first cigar, and then another. He became ill on the trip back to the fort.

He met girls from wealthy families who owned impressive vehicles, including a Cadillac limousine and a seven-passenger Haynes. He had a special relationship with a girl named Georgia, and he took many playful photos of her

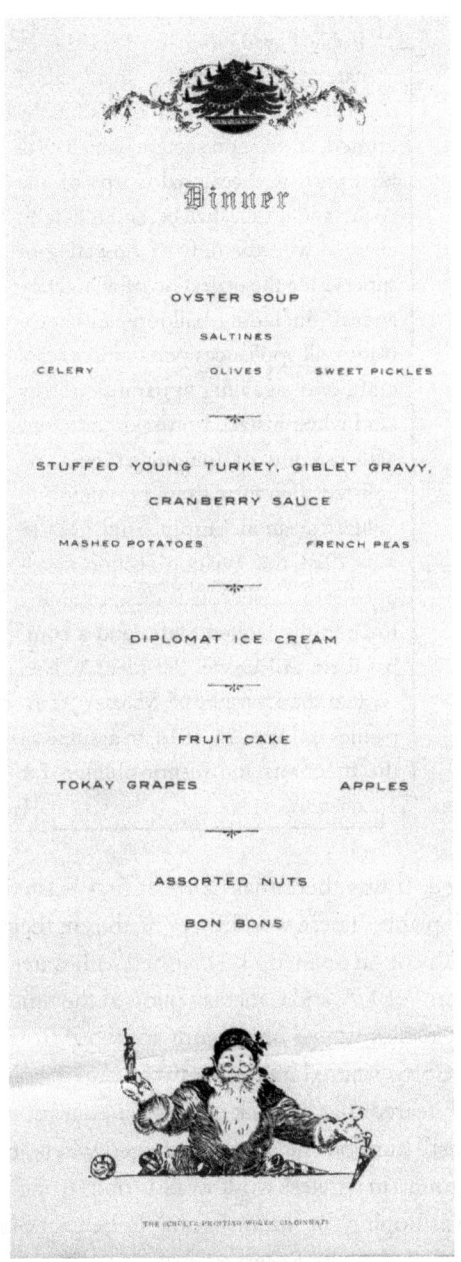

Soldiers at Fort Thomas were treated to a sumptuous Christmas meal that began with Oyster Soup and ended with Bon Bons. (Author's collection)

with his new Kodak camera. The letters he wrote to his family were frank and revealing, motivating Ada to send him some motherly advice: "Do not do her an injustice, Clarence. Act toward any girl as you would want any young man to act toward Irma." He answered that the girls were good and respectable and that he felt "mighty happy."[2] Ada also chastised him for using his modest income on weekly dates with Georgia. This too he defended, saying that the ten cents it cost to visit her and the money he spent on movies were not out of line.

By September, Cornish had learned that there was a branch of the Army in which he might be able to put his radio skills to use. He wrote to his family to let them know that he planned to apply for admittance to the Aviation Section of the U.S. Signal Corps, which was responsible for operating and equipping all of the military's planes and other aircraft. The news stirred Ada's deeper fears; concerned that her only son might get killed if he was accepted into an aviation field, she warned him against it. Nevertheless, the restless young man took the needed exam in the fall of 1917. "Whoops m'dear!" was his elated reaction when he learned that he had passed that first hurdle and now needed to report to the Examination Board at Wilbur Wright Field near Dayton, Ohio, for a personal interview.³

Cornish's brief stay in Dayton was a revelation to him. He wrote home about his excitement at seeing all those "aeroplanes" in the air. "There were about a dozen all the time. As soon as one would come down, one or two would go up."⁴ The field where he observed this activity had earlier been known as Huffman Prairie. It was the world's first flying field, the place where Wilbur and Orville Wright conducted early heavier-than-air experiments and established their flying school. The Signal Corps had recently acquired the historic site, and it was now part of the new Wilbur Wright Field, where the Army's pilots, armorers, and aircraft mechanics trained. It was there that Cornish first feasted his eyes on the Curtiss JN-4D Jenny biplanes. These workhorses of the air, their wings hand-built of wood, covered with cotton or linen, and brushed with waterproof dope, were used to train 95 percent of U.S. and Canadian pilots at the time.

> Public Law 63-143, July 18, 1914, states: "Be it enacted... that there shall hereafter be, and there is hereby created, an aviation section, which shall be a part of the Signal Corps of the Army, and which shall be, and is hereby, charged with the duty of operating or supervising the operation of all military aircraft, including balloons and aeroplanes, all appliances pertaining to said craft, and signaling apparatus of any kind when installed on said craft; also with the duty of training officers and enlisted men in matters pertaining to military aviation." From April 1917 to May 1918, the Aviation Section developed into parallel air forces, a training force in the United States and a combat force in Europe. President Wilson created the Division of Military Aeronautics on April 24, 1918, to assume all the functions and responsibilities for aviation.

After returning to Fort Thomas, Cornish waited impatiently for word from "up top" in Washington. The Signal Corps wanted men with two years of college or the equivalent, so he naturally feared that his lack of formal education beyond high school would "go against" him, but he was encouraged when an Army captain told him that his background in wireless work would "pull strong" for him. He soon had the news he was hoping for: he was going to be part of the military's aeronautical arm, the Signal Officers Reserve Corps (S.O.R.C.).

The United States' own Wright brothers had accomplished the first successful heavier-than-air flight, but the U.S. military did not initially recognize its potential. If they were aware of the prediction that a French Army officer,

Captain Ferdinand Ferber, had made in 1906 in a memorandum to the French War Department, they did not heed it. "The possession of a fleet of airplanes," Ferber wrote, "would give a general the possibility of knowing the moves and countermoves of the enemy.... No matter how quickly the enemy might mobilize his resources, not even the most rearward reserves would escape notice.... Under these conditions, victory is certain."[5] The argument that airplanes would revolutionize war was initially met with skepticism. As Ferber had predicted, however, once military leaders understood that airplanes were faster, more maneuverable, and easier to hangar than airships and dirigibles, they quickly responded. As a result, Europe was ready to fight in the air when war came in the summer of 1914.

On August 1, 1907, Captain Charles Chandler took charge of the newly established Aeronautical Division of the U.S. Signal Corps. He was responsible for "all matters pertaining to military ballooning, air machines, and all kindred subjects." A few months later, specifications were issued for "the construction of a flying machine supported entirely by the dynamic reaction of the atmosphere and having no gas bag." After the *Wright Flyer* completed a successful five-mile cross-country demonstration, flying from Fort Myer to Alexandria, Virginia, at an average speed of 42.583 miles per hour, the military concluded that the craft met specifications and ordered one. On August 2, 1909, the *Flyer* became America's first military plane, designated Signal Corps Aeroplane No. 1 (SC1) and identified in the *Washington Evening Star* as "Aeroplane No. 1, Heavier-than-air Division, United States Aerial Fleet." Lt. Benjamin Foulois, who served as Orville Wright's navigator on that demonstration flight, became America's first military pilot; in fact, he was the sole military pilot until April 1911. By late 1910, SC1 was worn out and ready for retirement. Although it was past time for a replacement to be ordered, a congressman sputtered, "Why all this fuss about planes for the Army? I thought we had one." Despite some scattered resistance, General James Allen ordered five planes and took delivery of the first of them, a Curtiss Model D, Signal Corps Aeroplane No. 2 (SC2), in April 1911. A pusher, as opposed to tractor, biplane with a 50 horsepower engine, was a "one-seater," meaning that a pilot would have to learn to fly without the reassurance of an instructor alongside. The Army's remaining airplane order was completed and delivered by the close of 1911.[6]

Between 1908 and 1914, the American government spent a paltry $500,000 on military aviation, whereas France, Germany, Russia, and Belgium spent a combined $54 million. In 1913, the United States had fewer than one hundred certified civilian and military pilots; worldwide, there were twenty-four hundred. France had embraced the Wright brothers' invention, and French engineers had quickly gone to work. They had improved the biplane, developed the monoplane, and earned virtually every flying record that was set between 1908 and 1914.[7]

In the early stages of the conflict, European planes were designed and used for observation. The "war in the air" began in earnest when small single-seaters were built specifically for battle. Capable of great speed and rapid climbing, these planes had machine guns mounted in front that were accurately timed to fire between the whirring blades of the propeller without destroying them. For surveillance over enemy lines, heavier biplanes carried the pilot and an observer with maps who helped direct artillery fire. The planes could carry two machine guns that turned in any direction. Later, large and powerful biplanes and triplanes, built for long-distance bombing raids, bombed enemy factories and submarine bases. Near the war's end, some of the bigger bombers were equipped with steel armor that protected them against machine-gun fire.[8]

Engineers were also working to improve what is known as an airplane's ratio of weight to power. Lighter aircraft are faster and more maneuverable, and they can travel farther on less fuel, so the goal was to minimize the weight of the engine and maximize the amount of horsepower that it could generate. In 1914, the ordinary airplane motor weighed 437 pounds and delivered 112 horsepower, for a weight-to-power ratio of almost 4:1. By the close of the war in 1918, the United States had developed the Liberty motor, which weighed only 1.8 pounds per horsepower and offered improved speed. The supercharger, another important invention, provided compressed air to the engines, making it possible to fly at up to forty thousand feet. Newly designed oxygen supply systems enabled pilots to breathe at these extreme elevations.[9]

In any military conflict, knowing what the enemy is up to and getting the word back to friendly forces is vital. Aerial surveillance was still in its infancy when the war began. Early flyers were not sure how airplanes would augment the process, but airmen took it upon themselves to discover the capabilities of their machines. Henry H. "Hap" Arnold and Thomas Milling experimented with air-to-ground communication by radio. Arnold, a student of the Wright brothers and later a general in the United States Air Force, recalled the primitive form of early surveillance: "Without radio [wireless] communications, the rapid delivery of intelligence still depended largely on horsemen. We, the airmen, were to jot down what we saw on brightly colored pieces of paper and drop the weighted paper to the ground, where a cavalryman, galloping hell for leather, would pick it up and take it back to the command post."[10] Wireless telegraphy was another major advancement in the practical use of airplanes by the military. And with sophisticated aerial photoreconnaissance, every foot of the enemy defense systems and lines could now be studied and analyzed. An elaborate layout of overlapping aerial photos could be mounted on a wall, giving the Allied forces an overview of the entire German defense system.[11]

By the middle of the war, from the English Channel to Switzerland, a relentless, deadly armed struggle slogged on in the muddy trenches and amid the tangle of barbed wire where machine guns mowed down Allied troops. Young flyers with a new perspective on conflict openly mocked the vast bureaucracy of a war fought in this traditional way. Aviators saw the sky as a superior battlefield and decried the parochialism that had initially held back aviation in some places. Rea Redifer describes a war scene in *Once upon a Canvas Sky*:

> The airman was more often than not separated from the mire a mile below by little more than a hope and a wish. He looked beneath him into one vast brown murk. To him, the trenches looked like a labyrinth, scrawled by a burned-out match. Swathes of rusted wire bled from fuzzed gray to rust for miles between the trenches, and glints of water from the million pock-mark craters of shell bursts reflected the color of the sky. Chalk lines of roads ran geometrically toward and away from the trenches, and broken, skeletal villages faded away to the horizon. . . . To a man in the trenches, an aerial dogfight looked like a dream-like aerial ballet. He heard the distant echo of engines high overhead as slowly they circled and separated, and then, suddenly, one would fall away, trailing a gentle billow of smoke, downward across the sky. The man on the ground would watch bemused a moment and then turn to more immediate things.[12]

The American Heritage History of Flight captures the likely mood of those valiant men of the air.

> From the beginning many noble themes have been woven into the fabric of flying: man against gravity, man against distance, man against time. But the basic theme of the early years was as old as the oldest flying legend: man against death. Each of the first flyers understood that the air was an alien element into which man ventured only at his peril. Far from shrinking from the challenge, they welcomed it. The played a dangerous

game for the excitement of it—and accepted without bitterness the consequences of losing.[13]

The horrific drama had been playing out in Europe for more than three years by the time Cornish enlisted in the Aviation Section of the U.S. Signal Corps.

On this side of the Atlantic Ocean, heroic but unachievable pledges were made when the United States joined the war effort in 1917. Caught up in the initial wartime fervor, politicians approved $640 million for aeronautical purposes in the first war bill and vowed to "darken the skies of Europe" with a vast armada of American planes.[14] The reality, however, was that the country had only six planes and one dirigible, with a few men receiving poor indoctrination on hastily improvised flying fields in New York, Texas, and California. Training in obsolescent Curtiss Jennies left American flyers unprepared for the more advanced airplanes, such as the Sopwith Camel, that they would soon have to fly in combat overseas. Most who received early advanced training did not survive their first two or three patrols in France. Of the initial 210 American pilots sent to the western front in Europe, 115 were lost—killed, wounded, taken prisoner, or released from military service because of illness or injury.[15] The average life expectancy for a fighter pilot was a mere three weeks when Cornish began his training. Undaunted by the grim statistics, he was in Columbus, Ohio, by the beginning of January 1918, ready to start ground school at Ohio State University, and presumably flight school at Wilbur Wright Field near Dayton.

Though he would soon draw $100 a month as a cadet, more than three times the pay he had earned as a private, Cornish quickly ran out of money because of the demands of his new status. He would have to cover the expenses of a uniform, costing in the $40 range, and personal gear, including a trunk. He asked his parents to send him $10, registered, special delivery, so that he could buy the trunk, suggesting that they sell his storage battery to cover half the needed amount. Always hesitant to ask for money because of his parents' perennially tight financial situation, he helped them out whenever possible. For instance, when he learned of his sister Irma's emergency surgery for appendicitis that January, he advised his folks to use his Liberty Bond, once it had been paid for, to help defray the doctor bills. A fellow trainee agreed to "stake" him to ten more dollars until payday to help buy arctics (winter boots), chevrons, a white hatband, leather puttees (leggings issued to officers and enlisted men), a Signal Corps button, and more.

Cornish and the others were gearing up and doing daily drills, but despite the wishes of the eager group of cadets to get started, extreme weather conditions—below-zero temperatures and snow that mid-winter of 1918—forced Ohio State University to close. The situation was also dire nationwide.

> The weather itself seemed to be operated in the interest of the Kaiser. . . . The demand for coal was seemingly illimitable; the capacity of the railroads to deliver it was crippled as never before in their history. In zero weather cities shivered, and the poor suffered cruelly for the lack of fuel. Hundreds of ships heavy laden with necessities for the hungry people of our allies lay helpless in American ports with empty bunkers. Great factories engaged in manufacturing munitions and other supplies vital to our armies were shut down for lack of power.[16]

The federal government had to take control of the railroads to try to remedy the situation.

For several days, Cornish and the others awaited their next assignments as best they could. Not where he wanted to be, in ground school applying his knowledge of radio telegraphy, at least Cornish was not in immediate danger. In a letter dated January 12, he told his parents that he had seen people with frozen noses and ears. He was grateful for the woolen socks that his Grandmother Cornish in Canada had knitted and sent to him, and he was putting them to good use, wearing them to bed along with his arctics, "as it gets ducedly cold at night when the steam is turned off and the transoms over the windows are opened."

By January 15, the rails were cleared sufficiently for trains to resume operation. The cadets left Columbus for new locations—half to the University of California in Berkeley, and the other half, Cornish included, to the University of Illinois in Champaign for training with the United States Army School of Military Aeronautics. These two universities were among eight that offered ground school, using the model of instruction developed by the Canadian arm of the Royal Flying Corps.[17] Cornish, eager to complete the course and graduate, was apprehensive about his future after he learned that of sixty-five "fellows" in an earlier group, thirty-four failed near the end of the eight-week course. The young men all yearned to be pilots but had to master the basics of several disciplines before they could be sent to one of the few dozen airfields used for primary and then advanced flight training. Through their course work they learned

Squadron H graduates of the U.S. School of Military Aeronautics gather for a photo on March 23, 1918, at the University of Illinois, Champaign. Cornish is in the front row, fourth from the right. (Author's collection)

about airplanes, engines, military law and courts-martial, the Lewis machine gun, the theory of aviation, magnetos and carburetors, maps, airplane and silhouettes (friendly and enemy). They were taught wireless means of communication and how to transmit words by Morse code, using a buzzer and a key, not a voice.[18] The expected rate for the trainees was ten words per minute, easy for Cornish who conducted "buzzy in the ears" classes at night for his fellow cadets. His amateur telegraphy and wireless radio experience in high school was helping to move him toward his future vocation.

Having taken and passed all the ground school subjects, Cornish graduated in March, two months after being admitted. From Illinois he went to Love Field in Dallas, where he was quartered in an old cattle barn at the fairgrounds and slept on a metal cot with no mattress. "When we got up in the morning, crisscross patterns were all over our backs."[19] He admired the wings that the commissioned officers wore on their uniforms, and he hoped he would someday have the same. However, for Cornish and the other cadets who were eager to apply what they had learned, time moved slowly. The military, having been caught unprepared when the United States entered the war the year before, was still rushing frantically to create airfields, have airplanes built, and recruit flight instructors. It was a major undertaking that could not be accomplished quickly.

To help fill their hours, some of the cadets practiced Morse code, sitting at a long table with a buzzer and earphones. On one occasion, an instructor demonstrated his ability to produce twenty-five words a minute, challenging the men to keep up with him. Cornish could not match that output, but he came very close, managing a little over twenty words. Suddenly the instructor said, "Lieutenant Cornish, I want you in the office." It was rare for a cadet to be invited into the office unless he was receiving a special assignment of some kind or was about to find himself in some kind of trouble. "I wondered what the hell I'd done," said Cornish. But there was no bad news forthcoming. He had stood out for his telegraphic agility, and the instructor was merely curious about him. He was asked whether he had operated a station before entering the military. "Yeah, call letters 9FB," he answered, only to learn that the instructor had transmitted under call letters 5CH, and they had actually radioed each other before the war. The two men were glad to now be meeting each other in person. When Cornish was asked what he was doing out there with all the novices, he replied, "I do what I'm told to do." Thereafter he was excused from the practice drills in telegraphy, and the two chatted about old and new times—and their shared passion, of course. Cornish had found a fellow amateur radio buff, and thus ended his "buzzy in the ears" practice down in Dallas.[20]

Cornish was impatient to move on, but he would be moving on into a dangerous activity. The U.S. Army was ill-prepared for the novel task of training pilots to fight in a world war. Hastily sited airfields carved out of farm fields often held treacherous pitfalls. Aircraft were frail and subject to mechanical failure. Even the bicycle-type wire wheels on the planes constituted a threat; grasses and weeds on the sod fields could become entangled in the spokes, wrecking the craft and killing or injuring the pilot. Trainers had minimal training and experience themselves. Factors such as these claimed a heavier toll than the enemy in the early months of World War I.

The U.S. military readied thirty-five airfields in 1917 and 1918 to provide the needed flight training for the eager young men. But learning to be a flyer was not without extreme risk: "To face the enemy in World War I, pilots first had to survive flight training."[21] Cornish left Texas for Park Field in Millington, Tennessee, where his big day arrived on May 6, 1918: "I finally got today what I have been waiting and working for the last 3 or 4 months," he wrote, "to fly under instruction." Would he be gripped by terror, as some were during their initial flight, or would he be among those who relished the sensation from their

This 1924 photo shows Cornish filling the gas tank of a Jenny, similar to one flown while training to fly. (Author's collection)

very first ascent? For Cornish, it was a twenty-minute "joy ride." The air that day was "puffy," he said, and "just like riding along in an automobile on a rough road . . . or more like a choppy lake in a canoe." It was an experience he would celebrate for the rest of his life.[22]

In the 1919 edition of *The Wonder Book of Aircraft for Boys and Girls*, this profile of a would-be airman appeared:

> Twenty minutes aloft, strung up between earth and sky, with the wind howling past and the never-ceasing throb of the engine in his ears, will teach him more than all the books that were ever written. . . . First, [there is] the matter of health. Rushing headlong through the air with more than the speed of an express train; racing through the clouds, thousands of feet above the peaks of mountains; driving down thousands of feet at a time, at two hundred miles an hour, puts the heart, the nerves, and the lungs under a very great strain. Also, the boy who wishes to be an airman must lead a clean life and be fond of the open air. He must be not too tall nor too short. Must not be prone to seasickness. Have good eyesight. Not too much on the heavy side. Above all, must have nerve and willpower.[23]

In 1918, Cornish fit the bill. Billy Brock was his first instructor, taking him up in a Curtiss JN-4D, a two-place open tandem-cockpit biplane powered by a Curtiss OX-5 90 horsepower engine.[24] Brock is known as an "Early Bird of Aviation," one of 598 pioneers who flew solo before December 17, 1916. He would go on to be a pioneer in long-distance and transoceanic flights. In 1927, he and Edwin Schlee of Detroit made a spectacular flight halfway around the world, from Newfoundland to Tokyo, covering 12,295 miles in eighteen days.

The Flying Jenny in which Cornish received instruction was not a dependable craft. The valves sometimes warped and led to engine failure, he recalled. In addition, "The landing site was hazardous—simply a large sod field with no runways. Many pilots died during crash landings." Cornish survived the hazards, and after four hours and twenty minutes of dual, in-the-air instruction, he made his first solo flight on May 27, 1918. In Charles Lindbergh's words after his first solo flight, also in a Jenny: "You are completely independent, hopelessly beyond help, entirely responsible, and terribly alone in space."[25] Cornish was more matter-of-fact when he wrote of that experience: "It sure felt good to take that plane into the air knowing that I didn't have an instructor to tell me what to do, and what not to. I didn't need him, for the flying officer and Mr. Brock, my instructor, said that my air work was splendid, especially my turns and glide in the field. In fact, the fellows all said that Mr. Brock was asking them all to notice my turns around the corners of the field."[26]

He was soon practicing mandated stunts, designed to teach pilots how to pull out of potentially fatal situations. "Those were the days before parachutes, too," he said. "If you made a serious error, you could end up dead and embarrassed."[27] Parachutes were not required until 1922. Before that, some young flyers found them unwieldy and regarded the pilots who used them as sissies; desk officers felt that if a pilot had a parachute, he might leave his plane when he should have stayed with his gun and machine and gone down fighting.[28]

> Some flyers were using parachutes, but they were not airplane pilots. In the United States, parachutes were integral to smoke ballooning, and spotters in observation balloons were equipped with chutes packed in leather tubes. According to *The American Heritage History of Flight*: "No Allied pilots and, until near the end of the war, few Germans carried them. The excuses were both cruel and stupid. It was claimed that parachutes were not sufficiently reliable to justify mass production and that if pilots had them, they might be tempted to bail out without a fight. As a result, numerous lives were lost unnecessarily. It was not unusual for airmen to jump from blazing planes, preferring quick deaths to being roasted alive."[29]

The Air Service, United States Army, was established on May 24, 1918, and rapidly began working to develop an efficient training program. Even in 1917, every pilot was required to perform aerobatics: to sideslip, loop, imitate a fall out of control, and demonstrate a dozen other maneuvers. Although most training airplanes were outfitted with dual controls, allowing the student to learn by first following the instructor's control movements, many instructors knew only marginally more than their students. There was no consideration given to instructor qualifications or motivation, and these men received little supervision. The aerobatics performed above Park Field and the surrounding countryside included a tailspin, a loop-the-loop into a wingover, and then a tight spiral. The loop-the-loop was less sensational than the others, according to Cornish. However, with the plane upside down, gasoline leaked from the tank and sprinkled his face, giving him the feeling of being "chloroformed."[30]

Besides actually flying the plane, Cornish received training in wireless telegraphy, aerial navigation, machine guns, motors, and photography. He practiced aerial reconnaissance in the Park Field area. Flying a Jenny with a 90 horsepower engine and given a post office map, he was told to go out and photograph a specific road intersection. "When there, flying about five-hundred feet above the target, you took a picture through a tin can with matching cross wires top and bottom and attached to the fuselage. Flying as slow as you could (you're down to a critical point), looking over the side of the aircraft, it was a pretty hazardous operation. Several students just spun in when the aircraft stalled. At five-hundred feet if you go into a spin, ninety-nine chances in a hundred you're not going to recover before you hit the ground. Doing that [aerial photography] once was enough for me."[31]

Five weeks later, on July 1, 1918, he made his first solo cross-country flight. His handwritten notes describe this and his second cross-country experience. Following a triangular course, he took off in a Jenny on the first leg—from Millington, north of Memphis, then a ninety-minute trip across the Mississippi River to Clarksdale Field in Mississippi, where he arrived with a little gas left in the tank. The flight, intended to take a few hours, instead took over two days. Getting to Clarksdale was trouble-free, but a missing motor (spark plug trouble) forced him down at Lula, Mississippi, and resulted in a flat tire as well. Once both were fixed, he took off but encountered engine trouble again, this time at Walls, Mississippi, where he once more "snagged a tire" on landing. With no way to fix the problems, he waited two days for a "relief ship" to take him back to his starting point, Park Field. The next day he embarked on his second cross-country solo flight, to Wynne and Marked Tree, Arkansas. This one was relatively uneventful. "A cylinder on the left side was weak and had a loose piston," but it was nothing serious.³²

There weren't any navigational aids at all and no charts. We'd be told how to get to a point in Mississippi. "Don't go over Memphis. Stay away from there. You go around. But when you've gone around Memphis, there are five railroad tracks. Now you count. That fifth one is the one you take down to Lula, Mississippi, where we're going." That's where the old iron compass came in. Only the trouble was that occasionally these birds would miss one or take this one instead of that one and they'd be in the boondocks, wouldn't know where in the hell they were. They'd have to land and find out where they were from the farmer, or somebody on the ground.³³

Having experienced only minor problems, Cornish escaped unscathed from training, and on July 13 he was promoted, by authority of Secretary of War W. P. Jernigan, to second lieutenant in the "(Military Aeronautics) Signal Corps, National Army."³⁴ He wrote a letter to the commanding officer to request a ten-day leave of absence, explaining that he had "no readily obtainable finances for purchasing [his] officers equipment" and wanted to go home to see personally "to the disposal of such property as will enable [him] to purchase this equipment." The furlough was granted—his first leave in the fifteen months he had spent

Then Second Lieutenant Cornish earned his wings in the Signal Corps. (Author's collection)

in the service. He visited his family and took care of financial matters, then boarded a train for John Dick Aviation Camp in Texas. After a brief stay there and a return trip to Dayton, he was southward-bound to Payne Field near West Point, Mississippi. Having become stranded near Prairie on the way to Payne Field, he sent a postcard to his family saying that the tender of the train engine had left the track and torn it up. Twenty miles from the field in a Podunk town with nothing but a couple of stores and a cotton mill, the troops were waiting for people at the field to send a bus for them.

At Payne Field, Cornish flew whenever the weather permitted, sometimes spending four to five hours a day aloft. One morning he went up to eight thousand feet. It took him half an hour to reach that altitude, but he lost twenty-five hundred feet in three or four seconds doing a tailspin. A few days later, he happened to be the only one who was up doing stunts. With the air to himself, he took the opportunity to do everything he could think of—barrel loops, reverse barrel loops, barrel rolls, and flying upside down. "You don't see any of it around here. In fact, I have never seen any of it around a flying field. I came down and found that the fellows, instructors and all, had been watching the exhibitions. I didn't do it for their benefit, or anything like that, but I was just having a good time." Flying had gotten under the showman's skin. "I don't know what I will do if I have to give up my flying after the war," he told his family in a letter. "I just naturally like the doggone stuff."

> Usually there were four or five of us up there at the same time playing around among the clouds and doing stunts as we felt like it. In other words, completely shut off from the ground below by a soft intangible mass of something white and—well I can't describe it—and just naturally enjoying ourselves. Up there that way, makes you feel that you are way off somewhere in a land far remote from this little globe of ours, and in the land of "fairies." It is as if the heavens opened up and we sailed into a new sphere. You sail down a ravine between two high banks of clouds, around and up a side cut in one of the banks and probably come out into a big valley with clouds for a ground, and higher and fleecy, flimsy ones above and the sun shining bright and serene on the whole scene. Then, again maybe, another plane will sail out into this valley from somewhere; you circle toward each other, wave a greeting and pass on. At a little distance apart one will do some stunt and then the other. Then after another greeting, sail on out of little peaceful valley to play around among the hills

and vales once more. These hours are made up of seconds never to be forgotten, and frame a picture of such wonderful, peaceful serenity, and such a picture that no one can begin to appreciate unless he has actually seen this enacted. But having seen and felt, I am greatly impressed by it. And so I have endeavored to put into a few colorless and insufficient words what I saw there above, yet among, the clouds yesterday.[35]

While Cornish and the other stateside flyers in the Air Service were biding their time and waiting for their next orders, Billy Mitchell, deputy commander of the Aviation Section and a brilliant tactician who commanded all American air combat units in France, called for a concentration of airpower over the battlefield at Saint-Mihiel, France.[36] Fifteen hundred airplanes (none of them built in the United States) belonging to the combined Allied air forces assembled to drive the Germans from the sky, seal off the rear, destroy communications, and aid the ground troops who were massing for the attack. Regarded as "singularly successful," the operation deflated the already weakened Germans, and impressively reinforced the effectiveness of military aeronautics.

In October 1918, the month before Armistice Day, Cornish rode the rails through Mobile, Alabama, and Jacksonville, Florida, to his final two posts near Arcadia in Desoto County.[37] There were two training fields in this part of Florida's flat heartland—Carlstrom, used for aerial gunnery, and Dorr, used for pursuit flying. They had been named in honor of Victor Carlstrom and Stephen Dorr, who were killed in separate aviation accidents during 1917. Eight miles separated these fields, where in 1917 the perimeters had been ditched, land cleared, and barracks and hangars hastily erected, the latter with concrete floors and doors on both ends. Hundreds of U.S. Air Service pilots trained there and won their wings. Unfortunately, several of those pilots died in crashes, and several others disappeared on training flights over unsurveyed prairies and the Everglades. Even into the late 1930s, pilots' remains were still being found in remote areas amid the wreckage of World War I Jennies.

Rumors were swirling that the war would be over any day now. Cornish feared that he might not complete training, and for a while, disappointment and frustration consumed him. To add to the malaise, everyone was confined to the base because of the Spanish influenza epidemic, meaning that there were no leisurely breaks in Arcadia. Mosquitos found their way under protective netting and left him sleep-deprived. "I'm thoroughly disgusted with the whole works," he said. "I don't seem to care a rip whether I fly any more or not. Darn poor grub too." Things did improve, however, and Cornish completed advanced training at Carlstrom.³⁸

Back in the air, his spirits lifted as he tested himself and improved his aviation skills, including a vertical reverse. "You start out by flying in a very tight circle on your side, you cut your motor, turn over on your tail, and while your nose goes past the vertical, give your throttle full, catch yourself and fly a circle on your side in the reverse direction. You do this stunt back and forth four or five times—supposedly without losing any altitude, or very little. . . . The first time I tried it, I lost something like 1,200 feet."³⁹ His final assignment was instructing other flyers in pursuit and in aerial gunnery, using Thomas-Morse Scouts and Jennies with Hispano-Suiza 150 and 180 horsepower engines. Pursuit pilots were trained to destroy hostile aircraft wherever they found them, to deny hostile aerial forces freedom of action, and to protect friendly operations against aerial attack. A mere four months after completing flight training in Tennessee, and with little further instruction, Cornish found himself teaching others the techniques. "I didn't know anything about fighting Germans in the air and neither did my students. It was the blind leading the blind," he recalled later.⁴⁰

Cornish wrote 125 letters to his family in Fort Wayne during his twenty-one months of military service. (Author's collection)

Cornish soon added a maneuver known as the Immelman turn to his pursuit tactics. Devised to allow a second attack on an enemy aircraft, it required precise control of the plane at a dangerously low speed. After making a high-speed diving attack, the attacker climbs back up past the enemy aircraft and, just short of the stall, applies full rudder to yaw the aircraft around. This puts the aircraft facing down at the enemy aircraft, making another high-speed diving pass possible. As he explained it, "First you do a half loop; then do a half roll and you've gained altitude, see. But you're going in the other direction. That's fine and dandy because often that puts you right in a position . . . to turn and dive down immediately on your opponent."

The plane he trained in was a Thomas-Morse Scout. A single-seater with enough space for the pilot only, it was powered by an 80 horsepower Le Rhône rotary engine, its propeller fastened directly to the crankcase. "The whole propeller and engine rotated around a stationary crankshaft," he explained. "You had two controls—one was gas and the other air. It was your business to make the mixture that gave you the best performance. They used castor oil in those engines because castor oil didn't mix with gasoline." One time several Thomas-Morse Scouts arrived at Dorr Field from a training field that had been inundated by heavy rains. When Cornish discovered that some of them had more powerful engines than his training plane, he took advantage of the increased horsepower.

> A lot had 110 horsepower Le Rhone [possibly Gnome] engines, but visually you couldn't tell the difference between an 80 horsepower engine and the one with 110. Well, I heard that there were a couple of those airplanes down in the hangar that weren't being used—the 110s. I got authorization, as instructor, to have two of them put into [good] condition—one for me and one for my buddy. We were instructing and combating with 110 horsepower engines; the students had the 80 horsepower engines. And you know, they never could figure out why in the hell Cornish and Greedy were always on top of them. Well, that 30 horsepower difference meant it was duck soup for us to get in the sun and dive on those birds.

The Immelman turn got Cornish in a bit of trouble one time. He had told his students, including the aviator and later aircraft manufacturer Eddie Stinson, to watch very closely. He flew to about four thousand feet over the field, did a wingover, and went back up to do an Immelman. After the wingover, he found himself in an inverted outside spin with full power on. Upside down, his hands left the joystick and his feet went up, though straps on the rudder bar kept his feet near where they needed to be. His fountain pen left his pocket and hit him on the way out. He saw the Pyrene fire extinguisher going up (but really down). He figured that he needed to cut the engine off.

> If I could get my hands down where I could get on that joy stick and cut the switch, at least I'd cut the power and my feet maybe get back on the rudder bar. I finally got that done. So help me I don't know to this day how I got out of that maneuver. I don't know what I did, but when I got out [of the situation] I was right over the outlying field and I skipped

> across the field at about 400 feet. Went around, landed, taxied up and about that time a rider on a motorcycle with a sidecar tore up to where I was telling the students, "Now that's not how to do it." But I didn't get a chance to tell them what happened. I was wanted at "stage." I climbed into the sidecar where they ate my butt out. . . . An inverted spin? An outside spin? Both prohibited!

He got down from the plane, kissed the ground, and told the lieutenant that he was thankful to still be alive. He lost his class and was grounded for seven days. "A sad experience, but I lived through it," he recalled much later.

In a letter home dated October 19, 1918, Cornish mentioned that he had flown in formation at eighty-five miles per hour while the lead pilot communicated with the others using a wireless telephone attachment, a novelty. Cornish's letters covered many facets of life in the military, so it is not a surprise that he closed his letter with an entirely different subject. "A fellow is showing the head of a huge rattler he shot this afternoon," he wrote in conclusion. "I guess I'll have to go and see it."

Aside from being home to giant rattlesnakes, Carlstrom Field offered a good location for ongoing experimentation. In the fall of 1919, future aviation luminary Lieutenant Jimmy Doolittle tested Charles Kettering's manless aerial torpedo there. Doolittle would later recall that after multiple failures, "a new type of altitude control was put on the last torpedo, constructed from salvaged parts of previous wrecks, and [it] flew about sixteen miles."[41] Despite what could be considered a success, he concluded, "The motor is not sufficiently reliable to permit safe passage of the torpedo over friendly troops." This device, known as the Kettering Aerial Torpedo (nicknamed the "Bug"), was a precursor to Hitler's buzz bombs launched against Britain and Belgium during World War II.

The Great War ended on November 11, 1918, with the signing of the Armistice at Compiegne, France. Thirteen million military personnel had died; twenty million were wounded, missing, or taken prisoner; twenty million returned unharmed. The League of Nations was established the next year, with a goal to reduce or abolish the likelihood of future wars. Sadly, that was a goal that it failed to achieve.

Despite the official end of combat, flying at American military airfields went on as usual. But in December, Cornish's classmates, men he had been with for more than a year, began to be discharged. He had not flown in battle,

Piloting this Thomas-Morse Scout was a far cry from what Cornish expected when he joined the U.S. Cavalry in 1917. (Author's collection)

and this was a dispiriting turn of events for him, though a fortunate one given the war's grim statistics for those who had made it to the front. With permission to do whatever he pleased in the air, he went aloft one last time as a military pilot, to give a thirty-five-consecutive-loop "performance."[42] A photo taken on the morning of January 7, 1919, shows Cornish standing in front of a Thomas-Morse Scout, the last plane he would fly in World War I. A few hours later, he was on his way home to Fort Wayne with an honorable discharge from the military. He was one of 8,689 American pilots who received preliminary flight training in the U.S. during the war, half the number of cadets who graduated from ground schools.

When the war began, the airplane was viewed by most people as "a dangerous toy for daredevils rather than a trustworthy vehicle for prudent travelers." Why would "ordinary human beings forsake the comfort, safety, regularity and relative rapidity of trains for city-to-city journeys and steamship for transoceanic voyages?"[43] With airplanes now recognized as much more than a "toy," foresighted Americans began to envision using aircraft to transport goods, mail, and people. It was a vision that would eventually become a reality—but not for many years.

3

Back to Earth

With the Great War now over, thousands of young pilots were left with few possibilities to put their training and experience to use. The interest in flight had been ignited, however, and the former military pilots were joined by civilians who were also looking for a way to take to the sky. For those who had $400 to spend, it was possible to buy a Jenny from the large fleet of excess U.S. Army Air Service planes. Some flyers eked out meager earnings as barnstormers, or by charging spectators at fairs and carnivals as much as a dollar to see an airplane up close. People who had never seen a plane in the air before now thrilled to observe the daring aerobatics and wing-walking. Although most among the general public would have been terrified to go aloft in an open two-seater plane, some brave souls with nerve and a few dollars climbed into the cockpit and became passengers.

Some pilots were driven to set records and gain fame. On February 27, 1920, Major R. W. Schroeder, the chief test pilot at McCook Field in Dayton, Ohio, tested his own physical limits along with the limits of his craft and his equipment when he climbed to 36,020 feet and, though short of his 40,000-foot goal, broke the altitude record. In the process he survived a drop in temperature to sixty-seven degrees below zero Fahrenheit and a failed oxygen system. Blinded by the subzero air temperatures and the speed of the falling plane, and barely conscious as his machine plummeted, Schroeder managed "within 15 seconds of the earth" to force open his eyelids with one gloved hand and grab for the controls with the other in time to effect a safe landing. The harrowing details were chronicled in the next day's *New York Times* and in the *Journal-Gazette*, which ran the story on March 5 under the headline "What It Is Like Seven Miles in Air." "Only one man, in the history of the earth, has ever been up there," where from that height he could see not only rivers and large landmarks but automobiles, streetcars, and cattle in fields.

Other veteran pilots carried parachutists aloft to compete for jump records. In 1921, in Rantoul, Illinois, Lieutenant Arthur G. Hamilton set a record of 23,000 feet. His record was broken the next year by Captain A. W. Stevens with a jump of 24,206 feet over McCook Field at Dayton, Ohio. In fact, every speed, distance, and duration record that was set by a U.S. aviator in 1922 was bettered by another American the next year. Even the military got into the competitive spirit when, in 1924, "round the world" flyers landed at Seattle after covering a distance of 26,345 miles in 175 days.[1]

Flying enthusiasts envisioned a future for civilian aviation, and they passionately promoted their cause. In June 1919, representatives from thirty-two countries attended the Second Pan-American Aeronautic Convention, held in New Jersey at the Atlantic City Airport. "Not less than 150 aeroplanes" were sold there, proving that there was indeed tremendous public interest. The need for American aeronautics development was becoming clear. And beyond the potential for efficient mail transport, practical applications were soon demonstrated. In 1921, when five thousand catalpa trees near Troy, Ohio, needed to be treated with insecticide to eradicate catalpa worms, it took "a mere fifteen minutes" to successfully spray them from an airplane. And in 1924, Army Air Service planes dropped bombs over the Platte River to clear an ice jam.[2] The public was beginning to realize that flying could be useful and perhaps even safe.

Nothing, however, lent itself better to the practical use of an airplane than the need to carry large quantities of mail from one part of the country to another. In the spring of 1918, even before the war's end, Congress authorized $100,000 for the first regularly scheduled airmail service, between New York City and Washington, D.C. A year later, a second route was added, this one between New York City and Cleveland. By 1920, U.S. Air Route No. 1, eventually known as the Main Line, crossed the country from New York City to Chicago to San Francisco.

Moving mail by air during those postwar years was novel, and its potential was compelling. But initially it was not the fastest transport method. Given the vagaries of flight, including the shortage of usable airfields and unpredictable weather that could cause delays and force unexpected landings, it was still faster to move the mail by rail. Moreover, flying the mail was a dangerous business, especially after dark, when there were few beacons for guidance. There were no radios in the planes, and open cockpits offered no protection from the elements. By 1921, nineteen of the original forty airmail pilots had died in crashes.

Facing criticism from the White House over the high cost and inefficiency of airmail, the Post Office Department instituted night flights to increase the speed of delivery:

> And so in the dead of the winter of 1921—the worst possible time—the Postal Service staged a bold and very nearly quixotic demonstration of fast transcontinental airmail delivery. Four single-engine, open-cockpit de Havilland biplanes took off, two from New York and two from San Francisco. One eastbound pilot was killed in a takeoff accident in Nevada, and the two westbound airplanes, halted by a snowstorm near Chicago, relinquished their cargo to trains.
>
> Pilot Jack Knight saved airmail. Meeting the surviving eastbound biplane at North Platte, Nebraska, Knight took off in darkness. Guided by bonfires and burning oil drums that had been lit by postal employees and helpful farmers, he flew all night in bitter cold, landing to refuel at Omaha and Iowa City, and reached Chicago in the morning. A relay pilot completed the trip to New York.
>
> Having covered eight hundred thirty miles in nine hours, Knight proved that the airmail could move even in darkness and bad weather. Lionized in the press as a hero (his name probably helped), he downplayed the difficulties he had faced, though he did concede that "if you ever want to worry your head, just try to find Iowa City on a dark night with a good snow and fog hanging around." The trail of fire that Knight followed was the beginning of the nation's first airway system.[3]

Soon Congress approved money for the Post Office to build airports and install beacons.

On February 2, 1925, Congress passed the Contract Air Mail Act. Called the Kelly Act after its sponsor, Pennsylvania congressman Clyde Kelly, it was designed "to encourage commercial aviation and to authorize the Postmaster General to contract for the mail service." The following year, the Air Commerce Act of 1926 authorized the secretary of commerce to establish airways for mail and passenger service, to organize air navigation, and to license both pilots and planes. As the government had regulated the maritime industry, it would play a similar role for aviation and "ports of the air."

There were increasing opportunities for flying in those immediate postwar years, but Cornish was unable to pursue any of them. He was an out-of-work flyer, one of more than eight thousand American men who had received preliminary flight training while in the service. He was figuratively but emphatically grounded; neither he nor his parents had the money he needed to get started in a civilian flying enterprise. While in the military, he had pondered the unimaginable—being confined to the ground after the supreme elation of soaring into the sky. What he dreaded had become his reality.

If he could not fly, he would need to work. A few days before his final departure from Carlstrom Field in Florida, he wrote home to his sister. He hoped that her employer might have a position for him: "Now here is where you can do something for your brother, Irma. I'll need a job right quick when I get back. So if you see any vacancies around the office that will pay at least a couple dollars a week, grab it for me. I'll be coming home with something like $125 in my pocket, but you know that won't last long." But apparently there were no openings that he could fill.

The next few years would test his mettle. From an average boy in a moderately sized Indiana town, he had rapidly evolved into a handsome cadet with regular pay who had survived the hazards of wartime flight training to become a skilled pilot. He was admired by young women, including one in Lakeland whose Cadillac 8 he drove one November day. He had danced in Tampa, Florida, with "the most beautiful girl he'd ever laid eyes on," a starlet who had a principal role in the play *Canary*. How would he cope with the sudden transition from independence and glamour to again living in his parents' home in Fort Wayne as his money dwindled?

"You couldn't buy a job when I came out of the Army in 1919," Cornish said later.[4] But he had to have work, and he found plenty of opportunities for short-term employment, including a job in April painting the Crawford, a hotel in Crawfordsville at the corner of Green and Main Streets, built in 1900 by the Jones family. In late spring, as Cornish was settling back into life as a Hoosier, the city of Fort Wayne organized an extravaganza to welcome back the returning war veterans. *Home Again*, held at the Palace Theatre on May 22, 23, and 24, featured songs, chorus lines, minstrels, and ballet performed by both veterans and civilians, including the "Snakey Siren of the Seine," Chester Hinton. The souvenir program was presented to the World War veterans of Allen County by the businessmen of Fort Wayne.

At some point during the challenging postwar years, Cornish worked briefly for his friend George Malone as a "picture man" in the hills of the southeastern United States; riding the train to small towns, he would get off and follow dirt roads, going door to door in the backwoods to sell framed pictures, samples of which he carried in a satchel. More often than not, the man of the house was away, so Cornish would make friends with the wife and entertain the kids in the yard before talking the woman into ordering a piece of art. She would either sign the agreement or, if illiterate, place her "X" on the bottom line. However, if the husband returned and found Cornish there, he would be run off. On a couple of occasions, he stumbled across bootleggers with their shotguns loaded. When challenged, he calmly reported that he was just heading over the hill. And they, in turn, allowed that he had "better just turn around an' go back where ya come from." At a later time, someone else, like Malone, showed up to deliver the framed picture and collect the promised money, the harder part of the sales job. Working as a "picture man" was not Cornish's higher calling. His legitimate jobs in Fort Wayne included collecting payments and making drawings of gas lines for the Fort Wayne Gas Company; clerking at the Bass Foundry and Machine Company, which supplied the railroads with steel train wheels; reading copy for the *News-Sentinel*; and working in sales for Holland Furnace Company.[5]

Living at his parents' home, with modest earnings but also modest expenses, Cornish found his off-work days relatively carefree. Shortly after his return from military service, a friend surprised him with a party in his honor: "Progressive bunco, a popular parlor game, was one of the main features. . . . Later a delightful luncheon was served. Those who participated were Misses Alma Scherer, Irma Cornish, Edith Baur, Alice Schmitt, Ruth Gallmeyer, Margaret Paulsen and Helen Gerberding and Messrs. Paul Sarver, Carl Strobel, Art Puff, Louis Scherer, Max Irmscher, Clarence Cornish and Ross Parnin."[6] In the fall, Mildred Koehlinger invited him to a Halloween masquerade party attended by thirty-four young women and men. She decorated her home in a black and yellow color scheme, with pumpkins, crepe paper, and cats. In one room, "music was furnished for dancing, while in another sat a witch in her tent ready to make known the fates of the guests." Once again, "a delightful luncheon was served."[7] His photo album from those years captures joy-filled excursions to Sylvan Lake, where friends, young men and women, swam, canoed, and picnicked on an island while listening to music on a portable Victrola.

Cornish was an active member of the Miami Athletic Club. At a club gathering in 1919, the members "enjoyed a banquet and social meeting at the Baltes Hotel. It was the first occasion, since the termination of the war that the membership has had a chance to get together. A majority of the members have served under the colors in some capacity and the evening was made a memorable one through each telling of his experiences while in the service."[8]

Cornish's Scouting spirit of adventure melded with his love for canoeing as he and his longtime friend Ross Parnin started out on a trip that he would never forget. "Two Fort Wayne Boys Shove Off on Maumee River," exclaimed the August 27, 1920, evening edition of the *Journal-Gazette*. The twenty-two-year-old "boys" had been best buddies in high school and were fellow aficionados of amateur radio. At 10:05 that Friday morning, they dug their paddles into the waters of the Maumee River at the Columbia Street bridge and began their dream trip from Fort Wayne to Canada by canoe, an eighteen-foot wooden craft that Cornish named "Kekionga" and that his father had hand painted.[9] The boat was loaded with the two "lads," each weighing 140 pounds, and 150 pounds of gear packed in duffel bags with bedding, foodstuffs, and cooking utensils for camping along the way, plus a shirt and tie, good trousers, and dress shoes for the planned return trip by train to Indiana from Toledo.[10] On the expedition, which they expected to take between ten and fourteen days, they planned to follow the river to Grand Rapids, Ohio, and the Wabash-Erie Canal to Toledo, edge their way near the shore of Lake Erie to the Grand River and Detroit, then cross to Windsor, Ontario, and finally reverse direction back to Toledo. The *Journal-Gazette* referred to the Toledo-to-Windsor portion as "perilous"; the *News and Sentinel* described it as "hazardous," to be attempted only under the most favorable conditions.

Good fortune was with Clarence and Ross as they began the journey in a river boosted two or three inches above normal by late summer rains. Headwinds on the 31st hindered their progress (as noted in a trip log prepared by Cornish), but only slightly. Fair weather generally accompanied the two. Photos taken along the way by both Cornish and Parnin capture the spirit of two happy-go-lucky young men having the time of their lives.

They did not go on beyond Toledo but returned to Grand Rapids, where they loaded their gear into Kekionga and bundled it in a thickly padded quilted wrap to ship home. Dressed in proper travel attire, they returned to Fort Wayne by train on September 4. They had realized part of their dream. More than half a century later, Cornish could still recall this carefree adventure in vivid detail.

During his early twenties, Cornish devoted much of his spare time to his old passion, delving into the intricacies and potential of amateur radio and operating station 9FB. As he recalled in 1975, he made receiving sets to earn some money, selling them to jewelers in Fort Wayne so that they could receive the time signals broadcast by a Washington, D.C., station and accurately set their clocks.[11]

In 1973, Cornish was the subject of an article in the *Indianapolis Star Magazine*. He talked about the early days of radio and recounted how he used to make "variometers (tuning devices)" by winding wire around an oatmeal box with a smaller coil rotating inside. He enjoyed experimenting with these units in various parts of his radio circuits. "About 1:30 AM I received a ham operator on a weak signal. I tuned

In 1920, Cornish's friend Ross Parnin took this photo during their attempt to reach Canada by canoe. (Author's collection)

the variometer and the signal built up, then fell off." Other adjustments brought in the strongest signal he had received, and he excitedly contacted friends, including his former teacher George Carter. "None of us could understand what was happening. But we went ahead making a better variometer—a ball rotating inside another ball—and improved reception even more."[12]

Cornish and his fellow enthusiasts were on the edge of rapid and vast technological changes in communications worldwide—going from wireless telegraphy, "King Spark," to radiotelegraphy using continuous-wave transmitters and vacuum tubes. "Beginning as an experimental application of new technologies for wireless voice transmission in the years leading up to the First World War, radio broadcasting in the U.S. underwent a period of dramatic transformation during the 1920s, developing into a new medium of mass entertainment that would forever alter the sounds and experiences of American life."[13] The first commercial radio broadcast was transmitted by Pittsburgh's KDKA in 1920. Suddenly it seemed that everyone wanted a broadcasting license, and because of the limited number of available frequencies, the crowded airways created chaos, making things especially difficult for amateurs like Cornish. Congress responded by passing the Radio Act of 1927, which eased the problem by giving the newly established Federal Radio Commission control over licensing, frequency assignments, and power levels. As the number of stations proliferated, the country was swept up in the "radio craze." Families gathered around their radio sets for entertainment and news. The entry of mass communication into American homes brought with it the development of a mass culture. "The same songs were heard across the country, news traveled fast, and heroes like Charles Lindbergh or Joe Louis were, in a new way, accessible to all."[14]

Cornish began to benefit from what he had learned in high school and practiced during and after the war. Though he and two others operated an active amateur radio station, 9FB, he also became part of the growing commercial radio industry when he and Verne Slagle opened the first retail radio store in Fort Wayne, the Wayne Radio Company, located at 1204 South Calhoun Street, in the front part of the Thor Electric Company store. The partnership with Slagle was short-lived; it dissolved after less than a year, and Cornish then became the sole owner of the business. The Wayne Radio Company had "enjoyed splendid growth as dealer in radio supplies and complete sets of established quality," reported the newspaper, and "Cornish would continue to supervise sales and the installation of radio outfits."[15] The stationery had a line reading *Wireless Telegraph and Telephone Apparatus* beneath the corporate name and stated that the company made and sold radio sets built from parts, was a dealer for A. H. Grebe and Company, and also did some broadcasting. That included furnishing the music for the first radio dance in Fort Wayne. The unique event took place at the First National Bank's

In the early 1920s, Cornish and Verne Slagle formed the Wayne Radio Company. (Author's collection)

annual banquet: "Young Folks Trip Light Fantastic to Music from Ether," declared a news headline. "Younger bank employees and some of the older ones . . . danced to a fox trot from Chicago, a one-step from New York, and a waltz from Indianapolis, while Denver favored them with an old fashioned quadrille."[16] Such broadcasts were forced to cease for "infringing on Armstrong patents."[17]

According to a 1924 ad for the company in the "Who's Who" section of the *Journal-Gazette*,

> RADIO Furnishes Amusement for All. The *tastes* of the average American are so varied that they can appreciate any part of the radio program. Buy a radio set today, and you will be entertained by just the music, lectures, opera, speech or sermon that will interest you the most. You will get the best results from the radio set and supplies bought here, The Wayne Radio Company.[18]

In that same section, the *Journal-Gazette* was offering $50 in prizes ($25 for first, $15 for second, and $10 for third) to the three people who, by 8 a.m. on March 6, could correctly identify the most caricatures of forty-two men holding a clue to their occupations. Among the men was Cornish, holding a rectangular box with four dials. In 1924 he was still identified primarily with his work with radios, not yet with airplanes.

Carter returned from Detroit in 1924 to give "one of the most comprehensive and thoroughly absorbing addresses about radio ever delivered in Fort Wayne," according to a newspaper article.[19] The former Fort Wayne High School teacher focused on the latest developments in wireless telephony. Immediately following his "electrifying" talk at the Chamber of Commerce, the radio enthusiasts who were present formed a local association. They elected E. R. Coolidge temporary chairman and Cornish secretary.

Fort Wayne's John L. Droegemeyer, with an acknowledgment to B. J. Duesler, wrote a "History of the Fort Wayne Radio Trade Association," in which he said that the informal group became official on a "miserable, cold, blustery, winter evening" in January 1925, when five radio store managers met in the back room of the Wayne Radio Company and formed the Fort Wayne Radio Trade Association.[20] That evening Cornish, whom Droegemeyer referred to as a pioneer radio dealer in the city, was elected the group's temporary chair. The local association was instrumental in organizing the Indiana State Association and was a member of the National Radio Trade Association, Droegemeyer reported. In

1927 the thirty-six members, "all the leading and most progressive jobbers and dealers in the city," held another of their successful radio shows, again at the Elks' Temple.

Professional life for Cornish was becoming increasingly rewarding on two fronts as the 1920s progressed. His radio business in Fort Wayne was growing, and he was enjoying a greater involvement with aviation, first through the reserves and then commercially. In March 1919 he had been appointed second lieutenant as a reserve in the Air Service; attached to the Fifth Corps Area, he attended a two-week reserve training session from July 16 to 30, 1922, at Godman Field at Camp Knox (now Fort Knox), Kentucky.²¹ On July 19 he was again in the cockpit of a Jenny, with dual instruction, and he soloed on July 28. After a three-and-a-half-year hiatus, he had returned to the air. He attended reserve flight training the next two years at Wilbur Wright Field.²² The 1924 session concluded on July 20 with a stag party at the Officers' Club from 4:00 to 8:00 p.m., followed by a farewell reception and dance at the post gymnasium at 8:30. The invitations included a humorous cartoon of a chaotic airfield scene showing bewildered flyers observing several upended planes with their propellers buried in the ground.

At the conclusion of U.S. Army Service reserve flight training at Wilbur Wright Field in Ohio in 1924, everyone was invited to a farewell reception. (Author's collection)

Cornish urged others to join him aloft by promoting enlistment in the Reserve Corps, Aviation Section, notably at an event held at Alden Field southwest of Fort Wayne on September 24, 1924.²³ Lieutenant James E. Parker, chief of the Air Service of the Eighty-Fourth Division at Fort Benjamin Harrison, Indianapolis, was in command of the visiting planes, which flew in formation over the field and city. One of them was piloted by Cornish; he and four others put on the exhibition and offered free ten-minute rides. Their purpose was to

A crowd gathered at Alden Field on September 24, 1924, to see Flying Jennies up close. The pilots were members of the Reserve Corps, Aviation Section, Fort Benjamin Harrison. (Author's collection)

generate support for a Fort Wayne flying field, to build up a flight unit of 195 men, to secure government planes for training purposes, and ultimately to establish airmail service to and from the city. To accommodate those who had not had a chance to fly during the scheduled times, Cornish was asked to stay in Fort Wayne an extra day before returning the plane to Fort Benjamin Harrison.[24] By the end of the event, forty men had signed their applications for enlistment in the service. If found to be physically fit, they would be administered the oath of allegiance in the Hall of Veterans of Foreign Wars two days later. Enlistees would be called to active duty for fifteen days a year and would be expected to serve in times of emergency by a special act of Congress. Members of Flight B could receive voluntary instruction in aviation and would get together occasionally at social meetings.[25]

Discounting the military bashes, the two-week refresher courses at Midwest military airfields helped to develop Cornish's flying skills, while meeting the challenges of myriad jobs helped to build his self-confidence over time. During the early 1920s, his organizational abilities grew along with his stature in the community. He enjoyed being involved in local and regional aviation-related matters, and he was always willing to play an active role. This was clearly evident when the case was being made for a municipal aviation field. In a comprehensive front-page article titled "Aviation in Fort Wayne" that appeared in the *Journal-Gazette* on January 18, 1925, Floyd Showalter, the owner of a local landing place and school called Flight "B" Field, named four local organizations that had formed committees composed of aviation-minded citizens who were tasked with considering the city's future in aviation, with an emphasis on developing a municipal airfield. Cornish served on two of these committees: one for the Exchange Club, with John Patton and Wayne Moellering; and the other for the Reserve Officers' Club, with Showalter and Robert R. Bartel. The accompanying illustrations included photos of Showalter, Cornish, and members of the Flight "B" 309th Observation Squadron; a map that showed Fort Wayne as a logical hub for flights in the Midwest; and a map that pinpointed local airfields at the time. Cornish later recalled some of what went on behind the scenes and the outcome:

> It was determined that there was no specific law on the Indiana Statutes to permit purchase of land for "airport" purposes, but there was one which authorized the city's Park Board to purchase land within a five mile radius of the city for park purposes. This was the law and was a basis for the ordinance to acquire 156 acres of abandoned farm land north of the city. At the final vote on the Ordinance, the council vote was tied and the then Mayor William J. Hosey cast the deciding vote in favor of the action. The land was subsequently purchased and Fort Wayne thus acquired [developed there] its first municipal airport which was officially dedicated on May [June] 25, 1925, and named Paul Frank Baer Field.[26] The legal problem involved in this case led to our sponsoring the adoption of the "Uniform Municipal Airport Act" by the Indiana Legislature and it became law by favorable action of the 1929 Legislature. It is still our basic municipal airport law with relatively few amendments.[27]

The proposal subsequently put before the City Council committee was for Fort Wayne to have "the finest aviation field in Indiana."[28] Fort Wayne's leaders wisely committed $38,400 to buy the abandoned farmland four nautical miles north of the city's center. Just in time for the dedication, "workmen completed leveling the field and cutting grass for the ascent and landing of planes. Two runways, each 1,500 feet long and a landing place 500 feet in dimension" were completed.[29] On June 25, 1925, the dedication featured stunts performed by the "greatest living ace," French aviator Captain Charles Nungesser.[30] Local air hero and war ace Paul Baer, for whom the field was named, was unable to

Our Fort Wayne Distributor—C. J. Goral—is an enthusiastic airman and photograph shows him with his new Curtis Oriole.

Cornish flew commercially for Goral Airways, a passenger service that was started in 1925 by the Goral Motor Company after its purchase of a Curtiss Oriole. Cornish is at center; company owner Clarence J. Goral may be at left. (Author's collection)

attend.[31] Working as a cameraman in 1925, he was "grinding out movie thrillers in Hollywood."[32]

That same year, Fort Wayne's Goral Motor Company, which sold Rickenbacker automobiles, formed Goral Airways. The company bought a Curtiss Oriole with a 180 horsepower Hispano-Suiza engine and hired Cornish to fly for them on a part-time basis. His first flight in the Oriole was memorable. "I took it out for a spin. The engine was heavy as hell, so the mechanic put an iron chain in the tail, trying to balance the silly thing."[33] Initially Goral hangared the plane at Showalter's flying field. Located on the Fred Buecker farm on Lima Road north of Fort Wayne, Flight "B" Field subsequently became Sweebrock Airport in 1926, owned by partners George Sweet and Paul Hobrock.[34] In 1928 the facility changed hands again, when it was purchased by Guy Means, a local General Motors auto dealer, and renamed Guy Means Airport. Other notable early pilots who flew from that base were Kenneth Siebert, Ford Fair, and Frank Lovejoy.[35] Cornish's personal entry into commercial aviation coincided with the dedication of Fort Wayne's first municipal airport, and this propitious timing helped propel him into his future profession.

Flying for Goral provided "many notable first experiences" for Cornish, most of which were well covered by the Fort Wayne press. One of these was a unique social event hosted by Mr. and Mrs. Matt Jones at Sweebrock Airport. "An innovation in delightful summer parties, exchanging a dip in the lake and a game of cards for a dip in the sky and a game of clouds. High scores were not counted and honors were distributed to all. Prizes awarded were a memory of a bird's-eye view of Fort Wayne from a height of several thousand feet." Cornish and Harold Preston were the pilots.[36]

Some flights were simple and straightforward, while others were complicated. In the early morning hours of September 3, 1925, near the town of Caldwell, Ohio, a thunderstorm ripped apart the *Shenandoah*, a Navy dirigible out of New Jersey. Thirteen crew members died. The *Chicago Tribune* chartered a Jenny to fly to Columbus, near the scene of the disaster, so that a reporter could pick up film from a photographer and then "dash back" to have photos ready for the next day's early morning edition. Bad luck struck. Engine failure forced the plane and crew down in Fort Wayne. The reporter talked Cornish into making the hop to Columbus and back while the mechanic repaired the engine.[37]

With the reporter and Cornish's mechanic in the front cockpit, they took off, following highways.

Captain Eddie V. Rickenbacker, one of only two American Army flyers (Frank Luce was the other) to receive the Congressional Medal of Honor in World War I, started the Rickenbacker Auto Company in 1920 and purchased the Indianapolis Motor Speedway in 1927. He was a visionary and an in-demand speaker, and some of the predictions that he made in 1919 were considered outlandish at the time. For example, he foresaw the use of "wireless-controlled compasses on all planes for land work such as going from New York to Chicago" and predicted that there would eventually be stations at Buffalo, Cleveland, and Toledo. "By special electric wave lengths these compasses will be controlled from these stations, keeping planes on their paths." Every city should have an airdrome, Rickenbacker declared, "as near the heart of town as possible and easily accessible by automobile and streetcar," or better yet, "landing fields on the roofs of buildings in the downtown districts." What most astonished his listeners was his prediction about overseas flights—that long-range airplanes would have regular schedules, New York to London, and "would be controlled by directional wireless from both sides in a similar manner to [over] land flights." He envisioned planes with "a string of kite parachutes that would be released as the machine is landing, making for a much softer impact on the ground" and planes that could maintain "a position over a given spot after the ship had taken the air."[38]

Before compasses were standard equipment, "you flew by the seat of your pants," said Cornish. They came to a Y in the road, and even though Cornish thought it was the wrong direction, dusk was approaching, so he followed the branch with the most automobile headlights. When he saw the lights of the town to which the road led, he knew that they were not near Columbus; the lit area was not large enough or bright enough. Running low on fuel, he had to land, but when he peered over the side of the plane, everything was dark. He made several passes to feel it out, then yelled to his passengers, "Here we go," and down they went, hoping to avoid trees, fences, and stumps. He had landed in a pasture, but "suddenly we were in the midst of a stampede. When we circled, the cattle bunched up in a corner of the field. When the strange 'bird' rolled toward them, they panicked—fortunately on both sides of the plane instead of over it," he related years later. "What's going on out there?" yelled a farmer with a shotgun from off to the side. He turned out to be "a grand guy" and sold them ten gallons of automobile gasoline.[39] The town they were near turned out to be Springfield. The field later became the Springfield Municipal Airport; Cornish attended the dedication ceremony.

Before leaving Fort Wayne, Cornish had called friends at Norton Field near Columbus and asked them to have cans of oil lit to define the landing area. Heading east, he and his passengers safely reached their destination, met a photographer, and made a beautiful flight back under a full moon with photos for the *Tribune*. There was one problem: the chartered Jenny was still grounded, so Cornish agreed to fly to Chicago. "There was an airmail field at Chicago with rotating beacons along with fixed lights every ten miles that guided airmail planes at night. After the flight would pass, the lights were turned off to save electricity. I figured I was ahead of the early mail plane and didn't expect any trouble, but all of a sudden I couldn't see any beacon lights." Fog had drifted in from Lake Michigan, obscuring his view of the ground. Cornish turned back to Goshen to pick up the lights again, but they were off. He had to "feel" his way toward Chicago, but the closest he came that night was Joliet, Illinois, where he made another forced landing. This one was easier because of the moonlight, but he was not as lucky: he "snagged" a tire. Another farmer with a shotgun leaned out his farmhouse window to assess the situation. He ended up driving the reporter to town to catch the train for Chicago. The next morning, with the tire fixed and the tank filled, Cornish flew to Chicago (possibly today's Midway Airport) and collected his $300 from the *Tribune*. "The pictures we brought back were spread across the front page."[40]

Even Charles Lindbergh encountered troubles in those pioneering days. While carrying airmail over Illinois in 1926, he experienced a mechanical malfunction that forced him to make an emergency landing in a field of new clover near Athens. His plane narrowly avoided flipping over when the wheels sank deep into softened soil. A farmer, Herman Dicks, drove Lindbergh into Springfield with the mail, where it was placed aboard a Chicago and Alton train bound for the Windy City. Two weeks before that incident, a parachute had saved Lindbergh's life when he was forced to bail out of his plane after running out of fuel while looking for a pathway through rain and fog.[41]

In his interview with Douglas Clanin, Cornish talked about some of his early flights, including the hazards of night flying and landing and how fields were marked for safety. He described the use of parachute flare clusters as primitive nighttime illumination devices. There were "three on the side of the airplane," he explained. "If you had to make an emergency landing, you'd fire one of those flares at cruising altitude, or wherever you were. That would light up the area, but it didn't last long. On the way down you'd kick another one out and then just before you land, you'd kick the third one out. That was an optional accessory, but most people weren't flying at night anyway."

The "runway lights" used for those rare nighttime landings were actually "old oil cans filled with refuse oil drainings. Pour a little gasoline on top and line them up to mark the landing strip. A torch switched them on. . . . Just those pots down there. And not too many of them either, but the closest thing there was to an instrument landing system. And in lieu of a control tower? An irate farmer with a shotgun offering emphatic suggestions as to how the pilot should proceed."[42]

One of Cornish's first flights with Goral had been to transport a nonviolent offender for the police department, but it did not go exactly as planned. En route from Flight "B" Field with Detective Sergeant Walter Kavanaugh on board for the thirty-mile trip to Paulding, Ohio, to pick up the man, a clogged gasoline supply line caused the plane's motor to stop while they were two thousand feet above the ground and forced an unanticipated, but safe, landing. It was too late in the day to continue, so Cornish and Kavanaugh returned to Fort Wayne by rail and went back the next day to make repairs and pick up the offender. This newsworthy event was reported by the *News-Sentinel*, which stated that for the first time in its history, the city of Fort Wayne had brought back a prisoner, a man who faced charges of spousal neglect, by airplane. Kavanaugh didn't worry that his prisoner would try to escape. "Any time he wanted to get out and walk, the atmosphere was right there waiting for him. You don't need any handcuffs for a prisoner in a plane."[43]

A mere three months after dedicating its airfield, Fort Wayne was eager and prepared to be part of a grand event that organizers in several states hoped would encourage the nation's aircraft industry to locate in the Midwest. Aircraft manufacturers at the time were located in Buffalo, New York; Teterboro, New Jersey; Detroit, Michigan; Dayton, Cleveland, and Troy, Ohio; St. Louis, Missouri; St. Paul, Minnesota; Omaha and Lincoln, Nebraska; Wichita, Kansas; Seattle, Washington; Oakland and Venice, California; and Wabash, Indiana, home to the Service Aviation Company, which built a commercial six-passenger biplane. On September 28, 1925, Fort Wayne welcomed "The Greatest Array of Immense Commercial Airplanes Ever Assembled." Seventeen planes taking part in the First Annual Aerial Reliability Tour (subsequently known as the Ford National Reliability Air Tour) had left Detroit that morning, landing without mishap at Fort Wayne's nearly completed Baer Field, one of several stops on the 1,900-mile course.[44] The *Journal-Gazette* praised the city's foresightedness "in providing an ample and convenient field for the accommodation of aircraft" as it and the country entered "a new and great era."[45] The editorial also praised Arthur F. Hall, the president and general manager of the Lincoln Life Insurance Company, whose dedication and perseverance had led to a flourishing local chapter of the National Aeronautic Association, which arranged for the visit of the planes and the tour's first stop.[46]

The planes, which arrived and took off in pairs over a three-and-a-half-hour period, flew on to Chicago and then Moline, Illinois; Des Moines, Iowa; and Omaha, Nebraska. Among them was a German-made Fokker, a ten-passenger monoplane powered by three Wright 200 horsepower air-cooled engines. The cabin was six feet high and luxuriously equipped, with a toilet on board. The Fokker and the Ford Stout Metal transport plane vied with German Junkers for the most public interest, but "the Fokker was easily the favorite for close examination and comment by the crowd," according to one report. "It resembled a mammoth bird, reposing on the line of inspection."[47] While visiting pilots were treated to lunch and their planes were being serviced, Cornish took Mayor William Hosey for a ride in the Goral Airways Curtiss Oriole biplane, brightly decorated in yellow.[48] Added a writer who had observed the assemblage of aircraft during their stop in Chicago:

> Soaring high in the air—2,000 feet up—a huge chicken hawk, silently attentive to what was going on below, watched the planes as they were making ready to take off yesterday noon. A number of spectators noticed the bird. He seemed to float with wings outspread on the wind with no effort at all. The bustle and whirr of propellers did not worry him; buoyant and serene, he drifted without a flap of wings. He appeared to know

a secret about the mastery of the air which the mechanical birdmen on the ground below had not yet learned.[49]

Fort Wayne's inclusion on the tour proved that city leaders had acted wisely in buying the land and equipment for Baer Field.[50]

There had not always been such collaboration at the state level. Prior to World War I, "patent litigation, personal jealousies and a lack of an appreciation of the true value of co-operation" had made working together to popularize the airplane and "convince people of its utilitarian value" a difficult process. However, "after entering World War I, the U.S. immediately realized the important role of aircraft in warfare." As a result, on February 13, 1917, the Aircraft Manufacturers' Association was formed. It "was enlisted to address specific problems associated with war-time production. The association drew up a cross-licensing agreement to allow manufacturers to have unrestrained use of airplane patents in order to produce airplanes for the government's war effort. The organization, later named the Manufacturers' Aircraft Association (MAA), continued to unify the air industry and engage in public education endeavors." It was subsequently dissolved, however, and in 1919 the Aeronautical Chamber of Commerce was founded, "with a charter membership of 100 'to foster, advance, promulgate and promote' aeronautics, and 'generally, to do every act and thing which may be necessary and proper for the advancement' of American aviation. Early members included such aviation pioneers as Orville Wright and Glen H. Curtiss, as well as representatives of major aircraft manufacturing units in the United States."[51]

In mid-October, Cornish was again mentioned by the *News-Sentinel* in its coverage of the sixth annual three-day "Get-Acquainted" goodwill tour held October 13–15, 1925. The Chamber of Commerce, which had organized the tours initially to benefit wholesalers and jobbers, expanded the event in 1925 to encompass all Fort Wayne businesses. "Not only those who might profit from closer commercial friendships between neighboring cities will go along with the party, but business men who are interested purely in giving Fort Wayne the important status to which it is entitled as second largest city in the state and the largest commercial center in northern Indiana."[52] That year the organizers added a bonus—an auxiliary air delivery flight. They arranged for Cornish, in collaboration with the *News-Sentinel,* to fly special "Good-Will" editions of the newspaper plus personal letters to the members of the party, who, carrying specially designed red, white, and blue umbrellas imprinted with "Fort Wayne

Sixth Annual Get-Acquainted Tour" and headed by their own band, paraded in thirty-seven towns and cities, including Columbia City, Plymouth, Peru, and Kokomo. Oscar G. Foellinger, president of the News Publishing Company, flew with Cornish and sat in the second seat of the open-air cockpit. Their ninety-mile flight to Kokomo took fifty-five minutes; they cut eight minutes out of that time on the return trip, the *News-Sentinel* noted.[53]

Sometime between 1925 and 1928, Cornish was the pilot on a special flight with a specific therapeutic purpose. Across the nation, word had spread within some circles that a sudden descent in an airplane could restore a person's hearing, either by shocking the nervous system or from the impact that changed air pressure at different altitudes exerted on the eardrums. "Expressing the belief that partial deafness is a condition which might be helped by a ride in an airplane, Mrs. E. W. Heckman, 822 Oakdale drive, recently accepted a ride with Clarence Cornish, air pilot. Mrs. Heckman imagined that she could hear better after she had alighted from the plane." She was "overjoyed to find she was able to hear her friends and her small son, Eugene, speak in ordinary tones of conversation" after a drop of nearly twelve thousand feet.[54] In the summer of 1925, Charles Lindbergh flew his first "deaf flight," in Lamar, Colorado, and received $50 for a single treatment. Lindbergh wrote to his mother that the day after that flight, his passenger's hearing was noticeably improved.[55]

There were celebratory flights, too. Mrs. Rachel F. Grodian's children gave her an airplane ride with Cornish in honor of her seventy-sixth "natal anniversary." A newspaper article, with an accompanying photo, noted that Mrs. Grodian, in her Sunday dress, pearls, and high heels, also wore a regulation helmet to keep her hair from blowing. She reportedly enjoyed the flight.

Mrs. Rachel Grodian is ready to celebrate her seventy-sixth birthday by flying with Cornish in an open-cockpit, bi-wing aircraft. (Author's collection)

And for Allen County citizens who wondered what to do on a summer day, Cornish encouraged them to join him with this notice:

> Make Your Week-end Complete
> See the Lake Country
> FROM THE AIR
> A trip over a few of the lakes
> of Noble county will do much
> to make your visit to the lakes
> one that will always be re-
> membered. Unless seen from
> the air the full beauty of these
> lakes is completely missed.
> FLYING OUT OF MAIL FIELD
> Sunday, July 25
> GORAL AIRWAYS
> C. F. CORNISH, Pilot
> 1st Lt. A. S. R.[56]

In 1926, Cornish made an experimental flight, circling over Fort Wayne while listening to messages broadcast by radio station WOWO. This was the first of several trial flights to test different radio equipment and determine whether transmitting equipment could be carried in a plane that would enable messages to be relayed from the pilot in the air directly to WOWO's "listeners-in." Cornish could hear the messages distinctly.[57] Captain W. S. Murphy of McCook Field in Dayton, Ohio, had conducted highly successful two-way experiments similar to the one planned by WOWO.[58]

Both radio and the aircraft industry were developing at a breakneck pace between 1925 and 1929. On May 21, 1927, Charles Lindbergh landed safely in Paris after a solo flight across the Atlantic Ocean from Long Island, New York. He became an immediate sensation and a world hero. The public was enthralled with airplanes, and in Indiana, as elsewhere, air events were hugely popular and attracted large crowds. On July 25, 1927, a headline in Peru's *Journal-Chronicle* proclaimed: "Thousands Thrilled by Air Races, Stunt Flying Here Sunday." Cornish was one of six pilots who participated in this two-day event, the article reported, which drew an estimated audience of twenty thousand from Lafayette, Indianapolis, Muncie, Fort Wayne, and other cities within the vicinity and from surrounding states. Cornish won the preliminary race with the Goral Curtiss Oriole and competed in the semifinals with two others. Jerry Marshall, the vice president of the Texas University School of Aeronautics, who was in charge of the air meet, described this "thrilling race" as the very best he had ever witnessed. Said the *Journal-Chronicle*, "The flyers were at almost all times flying very low in order to gain speed and at times seemed to be dangerously close together, except that Cornish, with his fast Oriole, was able to maintain almost a full lap lead at all times. So low did Marshall and Charlie Quinn fly when on the back third of the course that they were within three feet of the ground. Cornish prevailed with a record of 145 miles an hour." The three pilots divided the $250 purse.[59]

Marshall concluded Saturday's show with a nighttime pyrotechnical flight. Unbeknownst to the spectators, but as revealed two days later by the Peru newspaper, he suffered a near calamity.

> Beautiful as it was though, if the populace had known that the flyer, now at a dizzy height[,] was really dizzy and sick and worn from a strenuous day, and that his radiator was leaking, causing the temperature of his motor to mount to 205 degrees, thus threatening the total destruction of the ship in mid air, and that Jerry, mindful of conditions and with skillful control of the machine, caused it to dip from time to time in order to splash what little water he had left through the hot, and heating motor, only to temporarily reduce the temperature a measly five degrees, then they might have gotten a real for sure thrill out of the otherwise peaceful looking demonstration. Jerry told a *Journal-Chronicle* reporter, that if he had been in any other town but Peru, he would have abandoned the flight, but that he braved the situation in order that he might not disappoint the home folks.[60]

The experience illustrated the hazards that flyers risked then.

A month later, Cornish was part of a Fort Wayne–sponsored event, chairing the Field Committee for the first American Legion air circus. The *News-Sentinel* ran two editorial cartoons leading up to the August 27–28, 1927, event. The first, published on August 19, showed a bi-wing open-cockpit plane on the ground with *The Paul Baer Municipal Aviation Field* written on the side. Giant cobwebs adhered to its tail and a wing, suggesting a moribund air facility; placards surrounding the plane stated: "Wanted: Increased interest in local aviation"; "Let's show the world we can make it fly and fly high"; "Indifference and lack of support"; "Air Mail Route—National Prestige as an aviation center—Putting Fort Wayne on the air map." The cartoon strongly implied that there was a lack of support from the public for aviation and that promises had been made but not delivered upon. The same image was superimposed in the corner of an editorial cartoon that appeared six days later, but this time with the addition of a flyer enthusiastically waving his arms, promoting the air circus and urging, "Come on, folks! Let's go!"[61] If the cartoonist believed that the public had become jaded, he could not have been more wrong. Tens of thousands of spectators attended the affair, which was described in the Sunday morning edition of the *News-Sentinel* on August 28, 1927:

> If the late P. T. Barnum, dean of all circus-makers and author of the famous saying that the American people like to be humbugged, were living today, he would be the first to say that the American Legion air circus, which opened yesterday afternoon, was no humbug.
>
> There is nothing humbug about modern aviation. It is a science. And that science was well-displayed at the air meet which began yesterday afternoon and will close this afternoon before what is expected to be a record-breaking crowd at the field.
>
> Here is a circus whose big top tent is the blue sky and whose grandstand is as big as all outdoors. In the sweep and thrill of it all, one does not care that Bosco, the snake-eater; Roma, that strange girl, and the elephant act are missing.

The mesmerized crowd left reluctantly at the end of the day following an afternoon when "the sky was never free of planes which soared into the blue until they looked like so many swallows or grasshoppers" and "replete with races which would have made Ben-Hur's chariot seem like a crawling snail."[62] (Cornish took third place flying the Curtiss Oriole in a forty-eight-mile race.) The day's attendees took with them memories of dead-stick landings, formation flights and maneuvers, efficiency races, and model airplane demonstrations sponsored by the National Aeronautic Association. A *News-Sentinel* editorial on the 30th had nothing but praise for the air circus. Given the response by the public, the paper concluded that the city could no longer afford to neglect improvements to the municipal field, as "it manifestly represents a civic asset of tremendous value. . . . Hangars and better drainage facilities should be provided at the earliest possible date, and the roads leading to the airport should be paved without undue delay."[63]

Cornish, flying high in the mid-1920s, also had a steady girlfriend, Lois Lucile Watterson. The two had first caught one another's eye in 1920 as they passed going opposite directions on the sidewalk at the end of the workday. Lois worked as a stenographer for attorneys Levi Todd and Lee Hartzell and was walking to catch the streetcar that she rode home. During the following spring and summer, they saw each other almost daily. They had dates to ride the Whip, the Merry-Go-Round, and "everything" else at Trier Park. At Blue Lake near Churubusco, the girls went to the dance hall while the boys swam. As their years as a couple continued, they watched movies at the Jeff and the Strand, stopped for fresh cherry pie and ice cream at the Sunset, joined the senior Cornishes at Lake Wawasee over a Labor Day weekend, danced in Rome City to Fuller's Orchestra out of Chicago, attended a Ku Klux Klan parade, and watched a "Human Fly" climb the Keenan Building on the Fourth of July in 1923.[64]

Beginning in the early 1920s, Clarence and Lois Watterson became a twosome. They later married. (Author's collection)

Rome City and Sylvan Lake and their amenities often drew the couple there during their courtship. One Sunday in late May, Cornish loaded the canoe on his car and picked up Lois at 6:00 a.m. At the lake they paddled to a small park at the east end and sat in the shade, listening to music on a Victrola. The afternoon ended with a stroll on the grounds of the Gene Stratton-Porter house, and that evening, after a dance, they drifted around in the canoe for a couple of hours before returning to the cottage where they were staying with friends.

On December 4, 1923, Lois wrote dispassionately in her diary that Clarence "brought a diamond for me." The "diamond" was presumably an engagement ring, as he broached the subject of marriage in a letter written in

February 1924, asking Lois whether she thought she would be well enough for them to marry in May or June. How she answered his question is unknown. However, their courtship extended over an unusually long period because of Lois's multiple health problems. In her diary she detailed bouts with "Glandular Fever" and "Le Grippe," a knee problem that required electric treatment, and infected tonsils that had to be removed. In 1924, Lois's doctor said that she needed to spend two weeks in bed. Two weeks stretched to three months, during which time family and friends visited frequently and brought her bouquets of fresh flowers. "Nervous spells" continued into October, when, seven months after she had been sent to bed, her life regained a semblance of normalcy and she rode in Cornish's new Dodge coupe.

A few days later, she and her mother, Florence Watterson, went to Florida, where they stayed in the Mount Dora and Umatilla area until mid-May 1925. That was only one of several trips away for Lois; she traveled to both coasts, and often to Florida. The letters that Cornish sent to his "Sweetheart" during her long absences reveal emotions that he kept veiled later. He expressed his anticipation and growing ardor as their reunions neared. "You'd better put on your steel rib reinforcements as I feel that you might need them when I see you Sunday," he wrote. And "I've got so many kisses stored up I don't know what to do." Cornish was known as a highly competent man, in control of his world and upbeat, but in the spring of 1925, with Lois away, he expressed concern about his mental state: "My brain doesn't seem to function any more. I guess you will have to come back and take charge of me and things here at the store. . . . Perhaps if you were here, things wouldn't look so black."

He was also worried about his business. Occasionally there were weeks when radio sales were good, but more often than not, his letters referred to business as "rotten," with sales that barely covered expenses. It helped somewhat when he realized that he was not the only one hard hit. "Business in general is pretty poor just now around town," he wrote in April 1925. "The Pennsy shops have cut to a four-day, thirty-six hour week." And despite the apparent downturn generally, the radio business was growing, which meant that there was more competition. Then Wolf & Dessauer, Fort Wayne's highly respected department store, had a sale, marking down the price on radio sets. He bemoaned the prospect of ruined business for two or three weeks.[65]

Cornish wrote most of his letters to Lois late at night, before heading home from the store. January 28, 1925, was a good night for music: "About zero degrees, if not below, and a fairly clear sky." He had turned on a new Grebe

Lois and Clarence happily celebrate their wedding day, May 12, 1926. (Author's collection)

set and was listening to an orchestra play "some awfully mean jazz." That night, in a letter that was the most effusive and buoyant of the dozens he sent Lois between 1921 and 1925, he shared some news about the potential growth of aviation in Fort Wayne. He had attended a City Council meeting at which they had voted to purchase a flying field north of the city (later Paul Baer Municipal Airport), the one he favored, at a cost of $26,000. "Fort Wayne is finally on the aviation map. . . . We'll no doubt start action on the staging of some sort of an air circus this spring or summer. So we should have plenty of flying going on here for you to see when you get back to the old burg again." Having expressed that, in August 1925 he impressed his fiancée when he flew over her house in Goral's new Curtiss Oriole. "It is a beauty," she wrote in her diary. Three weeks later, on August 24, she took her first ride in an airplane.

Cornish and Lois remained true to one another despite their multiple separations and the stress he experienced in maintaining a radio business through difficult financial times and while holding a second job as a pilot for Goral Airways. In the spring of 1926, shortly after Lois returned again from Florida to Fort Wayne, they publicly announced their engagement. On May 12, 1926, they were married after a six-year courtship. Later in the evening, she summed up that first day of her marriage—a marriage that would last sixty-nine years:

> Wed. May 12. This is our wedding day. It is a perfect day with the sun shining beautifully. We were married at 10 A.M. with Rev. W. W. Wiant. The breakfast was served at 11 A.M. to Garnet, Charley, Ruth, Art, Esther, Aunt Dollie, Grandma, Mr. & Mrs. Cornish, Mother, Mr. & Mrs. Wiant. We had fruit cocktail, chicken in timbales, bridal salad and hot rolls, ice

cream in orchid & green and angel food cake, mints in orchid & green, with tall candles in orchid with green tulle. I wore a light green dress with tan georgette trim, silver footwear and my bouquet was of pink roses. We left with Ruth & Art [her sister and brother-in-law] about noon. Arrived So. Bend 5 P.M. Had a chow mein dinner at Ruth's. We have a room at the Hotel Statler.

The young couple had to watch every penny, as Cornish recalled in speaking of their wedding trip fifty years later.[66] "I had collected accounts receivable and bought a new set of tires. [After the wedding] Lois and I and the Reo touring car, with side curtains and everything, took off and wound up in South Bend the first night. From there we went on to Lansing, Michigan, visited Lois's Aunt and Uncle. We got back to Ft. Wayne with, I think, $1.27 in my pocket."

More than a year before their marriage, the couple had contemplated their future living arrangements. Clarence suggested that he might trade his car for an empty lot and have someone build a house for them. But he sounded a note of caution: there would be additional expenses involved, including "taxes, coal and furnishing an entire house." The elder Cornishes considered the idea impractical, he wrote in a letter to Lois, adding that his mother did not think that Lois was "well enough to undertake the responsibilities of a home." With an income of around $3,600 from the radio business, they were advised by Frank to rent for a year or two to save some money. Instead, Clarence left his parents' home on his wedding day, and he and Lois went to live at her mother's on Harrison Street.

Cornish was still running the Wayne Radio Shop at the time, supervising sales and the installation of radio outfits. He closed the business the following year, for reasons that are not known. He then became director, along with Herbert C. Wall and Richard W. Wustenfeld, and second vice-president of the Independent Supply Company, a wholesale firm that sold electrical, radio, and telephone goods with capital stock valued at $12,000.68. However, his vocational focus was shifting from radio to aviation as his reputation as a competent and responsible pilot continued to grow. In 1928, Ben Geyer and Arthur F. Hall, the owners of the Auto Electric Radio Equipment Company, named him manager of their newly created Aircraft Department. It was a career move that would mark a major turning point in his life. He would now be able to give full attention to aviation and its rapidly multiplying uses.

4

The Lure of the Skies

In 1928 there were a number of notable achievements in aviation: Amelia Earhart became the first woman to pilot a plane across the Atlantic; copilots Edward Schlee and William Brock, with whom Cornish had first flown in 1918, set a duration record in the air of fifty-nine hours, ten minutes, and fifteen seconds; Charles Lindbergh conducted a Goodwill Tour of Central and South America; Commander Richard Byrd and his crew flew over the South Pole for the first time; and Charles Holman set a record with 1,093 successive loop-the-loops in Minneapolis. Behind the scenes of such headline-grabbing feats, major technical advancements that had begun during World War I were occurring at a rapid pace. At the same time, intrepid individuals tested their own nerve and skills. "What a ripe plum aviation was [after World War I]," writes Walter Boyce in the *Smithsonian Book of Flight*, "full of records to be broken, first flights to be made, races to be won. It was a time when any individual could catapult himself (and, less frequently, herself) from the ranks of obscurity to the headlines—winning fame and fortune in a single flight. The war had created a need for heroes: aviation could provide."[1]

Cornish had already earned his own set of accolades, his own share of the "ripe plum," during the exuberant "Golden Twenties," and he would continue to do so. When Fort Wayne's Auto Electric and Radio Equipment Company decided to establish an aviation division, a fixed-base operator to be called Aereco, Cornish was recruited by the company's president, Matt Jones, to help organize it and serve as the manager and chief pilot.[2] An ad that ran in the *News-Sentinel* on March 8, 1928, stated that the company "will soon be an accredited sales agent for a well-known airplane manufacturer" and "will sponsor a flying school under the able direction of Lieutenant Clarence Cornish, Reserve Officer in the Pursuit and Gunnery Division of the United States Army Air Service. It will eventually establish flying service with Fort Wayne as its headquarters."[3] Aereco's newly formed division planned to handle freight and "to purchase a cabin plane for excursion services. . . . A school would offer courses in ground work, navigation, and meteorology for the purpose of actively training those who desire to qualify as pilots. . . . The classes will start in May." The article concluded, "Commercial flying is growing more popular every day throughout the country. Fort Wayne is now embracing this line of business, being headed by a company of responsible citizens."[4]

"Aviation is here," Aereco's promotional materials proclaimed. "It is the infant giant of the industries. Known two decades ago only as a dangerous toy, the present generation will see it full grown and accredited as the supreme transportive means of the age. . . . Courageous young men who want to break into this fascinating and alluring occupation of Aviation, or men who want to learn piloting from a sporting standpoint, can enroll now for a Pilot's Course."

A decade had passed since Cornish lamented that he might not ever fly again after the Great War. He had continued in the retail radio business and flown part-time for Goral Airways while providing for himself and his bride after their marriage in 1926. Now, having been chosen to develop and run the newly formed Aereco, he could finally devote his full attention and expertise to aviation. He would have a steady income in a full-time position doing what he had grown to love most and did best, flying. Initially he would pilot a Waco-10 with an OX-5 engine.

Aereco was an instant success. The company's first promotional event was a Boost Fort Wayne Aviation Day at Sweebrock Airport on July 1, 1928. It is fun to imagine what it was like to be there, hearing the metallic clatter of the engines, the singing of the structural wire and cables, and the whir of the propellers as ten planes flew over the city and zoomed past spectators at the field. Popular daredevil Gene Rock did a rope ladder act and a delayed parachute drop from a height of three thousand feet. And "when Miss Dorothy Curran, Fort Wayne air dare-devil, dangled from a plane at the height of more than 1,000 feet at the end of a parachute which failed to release, more than 5,000 spectators . . . were given a thrill and a demonstration of skill in piloting." She had leapt from the plane, which was piloted by Cornish, but the parachute, which

Cornish managed and served from 1928 to 1934 as chief pilot for Aereco, a fixed-base operator in Fort Wayne. (Author's collection)

was fastened to the side of the plane, had failed to release, and as the frightened spectators watched, "the girl hung at the end of the ropes, swinging back and forth in grave danger of being hopelessly entangled and crushed or dropped to death. Only the clever maneuvering by the pilot averted the tragedy when Cornish sent his plane into a spiral, and maneuvered to keep her free from the plane until the parachute released. For fully a minute the crowd stood breathless and white faced watching the short but intense struggle against death. Then suddenly the parachute became unhooked, the tangled ropes straightened and the huge umbrella filled with air, dropping the girl safely to the field." Afterward Miss Curran admitted that she had been frightened, but she was confident that "everything would work out all right."[5]

Shortly after its founding, Aereco received approval from the Aeronautics Branch of the Department of Commerce to conduct a flying school, the twenty-ninth in the country to receive the department's full approval, and the first in Indiana. It offered both ground school and flight training for those wanting to earn a private pilot's rating. It was a modest business at first, located at privately owned Guy Means Airport on Lima Road north of Fort Wayne. Its logo was a soaring bald eagle. The field was sod and the office bare. An unidentified newspaper clipping notes, "Lieut. Cornish has just completed the lining of the office of the Aereco Flying school. The wall and ceiling were lined with wall board and then painted in a light cream. A stove has also been installed to warm the red nose after coming down from the sky." The simple surroundings apparently did not deter would-be pilots. It was a busy time, and Aereco was

Three planes owned by Aereco are hangared at Guy Means Airport. (Author's collection)

able to add a Curtiss Robin with OX-5 engine in March 1929. In November of that year, another aircraft joined the fleet after Cornish made test flights in different models—a Cessna with a Wright J-6–7 engine, a Curtiss Thrush with a J-6 engine, and a Curtiss Robin with a Challenger engine. The latter, a high-wing monoplane, was chosen.[6] In 1930, Aereco's operations moved from Guy Means to the Paul Baer Municipal Airport. It had been upgraded to an approved flying school and was rated in everything except air transport. That spring a Fleet two-seat open-cockpit plane with a Kinner K-5 engine was added, and it would become the workhorse for training new pilots. Three months later, in July 1930, Aereco bought a Travel Air cabin monoplane with an OX-5 motor, which Cornish flew from the factory in Wichita, Kansas, to Fort Wayne by way of St. Louis and Indianapolis. That plane was equipped with "a wheel control and every comfort of the modern automobile."[7]

In February 1931, J. S. Marriott, chief of inspection services from the Aeronautics Branch of the Department of Commerce in Washington, D.C., notified Aereco that it had been designated an approved school for limited commercial pilots, one of only twenty nationwide. This meant that students were not required to log as many hours to receive the various licenses as they would be at an unapproved school. As Cornish had passed a "rigid test before inspectors" of the Aeronautics Branch, he was qualified to instruct students at Aereco under the new rating. To qualify as a school for flight instruction, all planes used for that purpose had to be licensed by the Department of Commerce, there could be no more than fifteen students per airplane, and the facility had to be equipped with a classroom for each one hundred students enrolled, with each classroom seating a minimum of twenty students. All flying instructors were required to be rated by the Department of Commerce.

In the Fleet training plane, trainees learned the vital skill of blind flying while under a hood.[8] Blind flying taught pilots to trust their instruments rather than their senses in low-visibility situations such as fog. With the hood completely covering the forward cockpit, the pilot could see the instruments on his dashboard, illuminated by two small lights, but not his surroundings, including the sky, the earth, and the horizon. On the dashboard were air-speed, bank-and-turn, and rate-of-climb indicators, a tachometer, a compass, an altimeter, a clock, oil temperature and pressure gauges, an altitude adjustment instrument, a starter button, and an engine primer. The student and the check pilot in the second cockpit communicated through a system of speaking tubes.[9] The student could be told the position of the plane and how to right it as needed. The safety pilot was in charge of takeoffs and landings; the student would take the controls at five hundred feet. At that point he was to make the plane climb gently until it reached one or two thousand feet, then maneuver the craft in gentle and steep turns, stalls, and spins. Should the trainee have the plane in a dangerous situation, the safety pilot would take control.[10]

During the more than six years that Cornish spent with Aereco, he wore many hats—events organizer, passenger pilot, instructor, promoter, speaker, teacher, salesman, daredevil, record setter. He organized, directed, and frequently participated in airfield and airport dedications and air circuses, as well as air parades over towns and cities, delayed parachute jumps, triangular course races, spot and

A blind-flying hood was used in flight training to teach pilots how to rely on their cockpit instruments for guidance. (Author's collection)

dead-stick landings, and stunts. He often competed in the events, and the newspapers detailed his achievements and winnings. He once earned $75 for placing second in a twenty-eight-mile race and $25 for placing fourth in a twenty-eight-mile free-for-all race in Defiance, Ohio.[11] And as if he were not already busy enough, he even organized an Airport Softball League for students and friends. On Monday and Friday evenings, the Aereco Aces and Aereco Dubs played at Baer Municipal Airport; on Wednesdays the games were at Guy Means.

Aereco offered rides in both cabin and open planes, with Sunday Specials costing $1.00, $2.00, and $3.00. (Author's collection)

Though taking passengers aloft was not lucrative, it was important in promoting the company. In good weather, "Sunday Specials" were offered for $1, $2, and $3 in both cabin and open planes, with two passengers able to fit into the front cockpit. Occasionally passengers were offered a flight for a penny a pound. The charge for longer rides could be as much as $5. Cornish may have been the one who composed the following for a promotional flyer:

> Listen! For $3.00 you can take a trip over the city—see its streets and parks spread out like a vari-colored map—its streets like the pattern of a checkerboard—its parks like patches of green velvet—its rivers like strips of gleaming bronze—its highways stretching away to blue-and-white horizons like silver ribbons stitched through fields that appear to be alternating squares of black and green plush—adjacent woodland copses given the appearance of little clumps of kale.[12]

One distinguished passenger was Santa Claus himself, who was photographed descending from Aereco's Waco airplane on Christmas Eve day:

> With no intent of making it hard on parents who may have been telling their children how Santa Claus will arrive tonight with reindeers and

descend through the family chimney, the staff photographer presents an exclusive picture of old Kris Kringle arriving here today at Guy Means airport in an airplane. Mr. Claus, of the Northland Clauses, stated to newspaper men upon his arrival that because of lack of snow and ice this Christmas, he decided to use the most modern of vehicles, the sky chariot.[13]

Aereco also provided charter flights. When an Anderson surgeon who was visiting Fort Wayne, Dr. Kenneth Ayers, received a phone call about a patient who had suffered a relapse, he chartered an airplane piloted by Cornish. In forty-five minutes, only an hour after receiving the news, the physician was back in Anderson at his patient's bedside. According to *News-Sentinel* columnist Bob Schott, Cornish was also the pilot on the longest chartered flight to date from any of Fort Wayne's airports when he took Richard A. O. Conner, the president of Magnavox, and Martin J. Wolf, a marketing advisor, on a business trip to Tampa, Florida. The three departed on October 9, the day on which they made the longest nonstop leg, seven hours and fifty minutes between Cincinnati and Jacksonville.[14] The Patterson-Fletcher Company, a prominent men's and boys' clothing store at Harrison and Wayne Streets, chartered a Curtiss-Wright Junior training plane from Aereco for the Patterson-Fletcher Flying Club, which at one time had 150 members, 60 of whom were flying members. Those who wished to take instruction from Cornish and Gene Campbell in the new plane had to be at least sixteen years old and have written permission from their parents. The aircraft, a pusher-type monoplane with its propeller mounted on the trailing edge of the single wing, was considered to be both easy and economical to operate. Cornish flew it from St. Louis to Fort Wayne, where it was disassembled and reassembled on the second floor of the Patterson-Fletcher building to be viewed by the public.[15]

Early flyers and students often met on Sunday mornings for airfests. Gathered for a photo of the group in 1931 were Cornish, Howard Mackin, Harold Johnson, Andy Canatsey, Walker Winslow, Matt Jones, Bernard M. Lloyd, James Fergerson, Howard Rohr, Bernard J. "Bud" Ringle, Ralph Walters, Eugene W. Campbell, Harry A. Offutt, Orville D. Louthan, Robert T. Schott, John Henninger, Ed Johnson, and Earl Johnson. Aereco also hosted dinners and brought in speakers to help promote camaraderie among the men. The meal served on December 29, 1932, in Hangar No. 2 at Paul Baer Municipal Airport featured rabbit as the main course. In 1934, Colonel Weir Cook of Indianapolis joined the sixty students attending the dinner party and spoke about "The Progress

The camaraderie enjoyed by early flyers is evident in this 1931 photo of a typical Sunday morning gathering. (Author's collection)

of Aviation." Films about parachuting, obtained from the U.S. Army Air Corps at Wright Field, were shown.

However, no matter how well trained those in Fort Wayne's flying fraternity were, accidents happened. So when a TWA pilot en route on a westbound flight from New York City to Chicago landed at Baer and reported that he had seen a plane spin to the ground in flames forty-five miles away at Delphos, Ohio, Cornish reacted quickly. He knew that two hometown pilots were scheduled to be flying passengers at the county fair there, and he feared the worst. After a hasty phone call to officials in Delphos, he took off in the *News-Sentinel*'s aircraft *Yankee Clipper* to aid his fellow flyers. Cornish must have been overjoyed when he saw Lloyd Pierce's plane flying overhead as he arrived. What had occurred was not an accident but a smoke balloon exhibition. The balloon pilot had deliberately untethered himself from the muslin envelope and parachuted safely to the ground. All the TWA pilot had seen was the unmanned balloon spewing oily black smoke as it slowly descended to earth; from his vantage point, he imagined an air disaster.[16]

Aereco sponsored annual meets "to give the sportsman pilot a chance to try his skill in competition and to encourage sport flying." The first and second such events were held at Guy Means Airfield, with contests for Class A pilots, who had more than twenty-five hours of solo flying, and Class B pilots, who had less. At the third annual meet, held at Paul Baer Municipal Airport in

1932, "Rosco the dummy parachute jumper of Aereco, was pushed out of a plane at an altitude of 1,000 feet, but his parachute did not open, as the rope tied to his releases ring broke as he was forced out of the plane. Rosco came tumbling down to earth only to split his flying suit and to show the rags with which he was stuffed."[17] It was noted that fourteen flyers, not including Rosco, took part that year.

Cornish promoted aviation whenever and wherever he could. He lectured at Tri-State College in Angola as part of its aeronautical engineering course; addressed the Engineers' Club at the University of Notre Dame about modern aids to aerial navigation; taught meteorology and aerial navigation through the Army School at Central High School; spoke to the Young Men's Republican Club about national air defense; and addressed the Exchange Club, highlighting the examination by the Department of Commerce Aeronautics Branch of motors, propellers, and other plane parts that would contribute to the safety of flying, opening and developing airways, and instituting cooperation with the U.S. Weather Bureau.[18] On April 11, 1929, he was featured in an article in the *Waterloo* (Ind.) *Press*. He had told the Waterloo Lions Club that the government had formed the Aeronautics Branch in 1926 in order to safeguard the public, and that licensing requirements finally existed. Planes must be airworthy, he said, and must withstand careful tests. He cautioned club members to ride only with licensed pilots in licensed planes.[19] (Cornish held Transport Pilot license no. 2128.)

Through his speeches and other promotional activities, Cornish became the acknowledged voice of flight in northeastern Indiana and beyond; according to one account, "Lieutenant Cornish, the youngest army aviator to serve during the World war," was "one of the best informed aeronautical authorities in the central west."[20] Widely regarded as a prominent member of Indiana's fraternity of early pilots, he got to meet and sometimes collaborate with some of the world's great aviators—Jimmy Doolittle, Roscoe Turner, Jimmy Haizlip, Eddie Rickenbacker, Weir Cook, Herbert O. Fisher, Michael C. (Mike) Murphy, and even Amelia Earhart. He also welcomed a number of prominent aviators to Fort Wayne. Ed "Red" Jackson, for instance, spent several hours as Cornish's guest at the airport, having flown from New York in a Curtiss Robin. Red may have been best known as the holder of the world's barrel roll record, with 483 consecutive rolls, but professionally he was the chief pilot for the Curtiss-Robertson Airplane Manufacturing Company.

On July 13, 1928, Sir Hubert Wilkins and Carl Ben Eielson landed at the municipal airport. They had been invited to Fort Wayne by the local chapter of the

Amelia Earhart, shown here in March 1935, was one of many aviation celebrities whom Cornish met over the course of his career. He stands here to her right; to her left are E. Ross McCulloch, president of the Board of Aviation Commissioners, and Mayor Harry W. Baals. ("Famous Aviatrix Visits Fort Wayne," *News-Sentinel*, March 21, 1935. Used with permission from the [Fort Wayne, Ind.] *News-Sentinel*.)

National Aeronautic Association to speak and show a film about Wilkins's flight from Point Barrow, Alaska, across the Arctic Ocean to Spitsbergen, Norway. The *News-Sentinel* published a detailed account of the pair's perilous undertaking, which, according to Wilkins, had required "patience and courage." It was these "necessary factors," he said, that had made their accomplishments possible.[21]

When famed aviator Wiley Post visited Fort Wayne, Cornish was one of the members of the reception committee who worked with the Chamber of Commerce to organize a parade and a special evening event in his honor. Post, with navigator Harold Gatty, had encircled the globe in eight days and fifteen hours, beginning June 23 and ending July 1, 1931. He landed at Paul Baer Field in the *Winnie Mae*, the ship in which he had made that famous flight.[22]

On August 23, 1931, Cornish, piloting a Ryan B-5, escorted Captain Lewis A. Yancey on his flight in a Pitcairn Autogiro from Lake Wawasee, in Kosciusko County, to Fort Wayne. They landed at Paul Baer Field shortly after noon. Cornish must have considered it a privilege to be chosen, as Yancey was the first transatlantic flyer ever to visit Fort Wayne.[23] Taking off from Old Orchard Beach, Maine, in July 1929, he and copilot Roger A. Williams had crossed the Atlantic in *Pathfinder*, a Bellanca monoplane with a Wright Whirlwind engine, arriving two days later in Rome, where the "winged messengers of Italo-American friendship" received a tumultuous welcome from a crowd of more than 100,000.[24]

The Pitcairn Autogiro that Yancey flew to Fort Wayne was named *Miss Champion*, and the strange-looking rotating-wing aircraft was a striking contrast to *Pathfinder*. Cosponsored by R. M. Kaough & Company and the Auto Electric and Equipment Company, *Miss Champion* served as a flying billboard for its owner, the Champion Spark Plug Company. Yancey treated passengers to rides in the rotorcraft all afternoon, taking off from both the Paul Baer Municipal and Guy Means airfields. The *News-Sentinel* published a list of Yancey's passengers; among them were Clarence and Lois Cornish. Before calling it a day, Yancey did an exhibition descent match with parachute jumper Gene Rock. "Rock leaped from a plane piloted by O. B. McVey at a height of 2,000 feet and at the same time Captain Yancey cut the motor of the 'windmill' type plane. Rock landed first but officials declared the descent speed of the two was about the same."[25] After spending the night in Fort Wayne, the famous pilot continued on his way to Toledo and then the 1931 National Air Races in Cleveland.

Autogiro pilot Lewis Yancey and parachute jumper Gene Rock compare their respective rates of fall in 1931. (Author's collection)

Yancey, in the center wearing white knickers, gathers with local officials next to *Miss Champion*; Cornish, who escorted Yancey on his flight into Fort Wayne, is fourth from the left. (Author's collection)

> The autogiro was invented by Juan de la Cierva, a Spanish aeronautical engineer. A precursor of the helicopter, the plane "sported stubby fixed wings and a front-mounted engine, but it had a large, unpowered rotor, which turned in the craft's slipstream and yielded additional lift." It "could fly with a short takeoff run and land nearly vertically, its rotor whirling like a pinwheel as it descended." Aviation pioneer Harold F. Pitcairn partnered with Cierva to build autogiros in the United States; "the first aircraft manufactured by the Pitcairn-Cierva Autogiro Company of America flew in October 1929."[26] Amelia Earhart praised its safety and ease of control after flying one and established an autogiro altitude record of 18,415 feet in a Whirlwind-powered Pitcairn at Willow Grove, Pennsylvania.[27] Cornish would be a proponent of a later development, the helicopter.

During a speech given in 1930 or 1931, Cornish highly praised American businesswomen for being more air-minded than their male counterparts, saying that they displayed little reluctance or hesitation about air travel. Women were enthusiastic users of the modern means of transportation, he said, and appreciated the airplane's speed and reliability for business-related trips.[28] Increasing numbers of women were also becoming pilots. According to the Museum of Women Pilots:

> By 1929 there were over 100 American women and numerous women in other countries licensed to fly. A Women's Air Derby was launched that year in conjunction with the Cleveland Air Races[,] which the women were not allowed to enter. The rules were straightforward. Whoever got to Cleveland in the least time won! The starting point was Santa Monica across the continent to Cleveland. At takeoff Will Rogers remarked that the start looked like a "powder puff derby." Louise Thaden claimed first in the faster aircraft class with her Travel Air J-5, and Phoebe Omlie was first in the second division.
>
> Despite the competitive nature of these talented women, or probably because of it[,] they felt their camaraderie called for a more formalized bond. After all had arrived in Cleveland, Amelia Earhart, Gladys O'Donnell, Ruth Nichols, Blanche Noyes, Phoebe Omlie and Louise Thaden gathered under the grandstand and, at the suggestion of Phoebe Omlie, considered forming some kind of organization just for women pilots. An organizational letter went out to all the licensed female pilots in the United States. Of the 117 licensed women pilots in the country, 86 responded to the call.

On Nov. 2, 1929, 26 women gathered at Curtiss Airport, Valley Stream, NY. The women conducted their business in a hangar as the work of the mechanics proceeded around them. Tea was served from a tool box wagon on wheels. Club eligibility and purpose were quickly decided upon. Membership would be open to any woman with a pilot's license, and the purpose was "good fellowship, jobs and a central office and files on women in aviation."

Choosing a name was a little harder. Amelia Earhart and Jean Davis Hoyt proposed the name be taken from the sum total of charter members. Thus the group was momentarily the 86s, then the 97s and finally the Ninety-Nines. The name/number stopped at 99, but the membership thereafter grew worldwide.[29]

Cornish gave generously of his time to promote aviation. Following a talk on World War I aces that he gave to a group of reserve officers, a memo dated March 7, 1933, was sent by a second lieutenant in the Cavalry Reserve to the commanding general of the Indiana Military Area in Indianapolis:

a. Capt. Cornish digressed sufficiently to give everyone present an idea of the function and use made of the various types of aviation by the American forces during the World War, so that a better idea was gleaned concerning aviation's war use.
b. Today Capt. Cornish spoke before the Athletic Association of the International Harvester Motor Truck Plant in Ft. Wayne regarding aviation of today, commercially, for pleasure, and military.
c. Credit should be given Capt. Cornish for doing more for aviation during the past year in Ft. Wayne by means of dozens of talks before all manner of business men's, young men's, athletic, and other forms of gatherings than all other men together in this Ft. Wayne sector over a period of several years.[30]

When Cornish was not transporting passengers, welcoming notables, or giving speeches, he enjoyed doing stunts. "In a way we did some barnstorming," he told a reporter in 1973, "but not like the itinerant barnstormer who flew from field to field all over the country. I was a fixed-base operator. We put on air shows for the Legion and other groups and did parachute jumps and aerobatics to attract people to the field for passenger hops."[31]

Aviation pioneer Jimmy Doolittle attended several events in Fort Wayne over the years. Left to right: Paul C. Guild, Cornish, Major Doolittle, Lieutenant Haizlip, Captain Robert Bartel, and Ford Collins, with Aereco's mascot, Pittsburgh, in the foreground. (Unidentified newspaper photo)

Cornish had a major role in the first of a series of air shows during the summer of 1934, presented at Paul Baer Municipal Airport under the auspices of the Board of Aviation Commissioners. Major Jimmy Doolittle, the honored speaker, first led an air parade over the city before beginning the program back at Baer with recollections of his active flying days in South America. He also outlined plans for the upcoming MacRobertson Air Race, to be held in October 1934. Also called the London-Melbourne Air Race, it was an event in which many of the world's premier pilots would be competing. The newspaper noted that Doolittle's voice was "amplified over a public address system." Then the celebrity headed back to St. Louis in his Lockheed Orion, and it was Cornish's turn in the spotlight. "The feature performance of the program was the 'dead stick' work of Capt. Clarence Cornish, manager of the Aereco Flying Service. Captain Cornish flew to an altitude of about 5,000 feet, cut out his engine, dived to gain speed, looped, dived and looped again, dived and executed a barrel roll and then landed, with engine still dead."[32]

"A good pilot is like a good bowler or a good poet—born and not made."[33] That was an apt description of Cornish. Bob McComb, a pilot trainee in Fort Wayne in the 1930s, described him as a consummate flyer with just the right touch in the cockpit. Mary Borton, Aereco's secretary, took part of her salary in flying lessons, and McComb, a man with excellent recall, later shared her description of flying with Cornish in a Ryan B-5. What left the greatest impression on her was his smooth handling of the craft, she told McComb.

> He leveled off from the fast climb in such [a] smooth manner that I didn't feel the least bit of sinking sensation. His smooth adjustments with the controls to set the plane on its course, deft use of trim tabs, adjusting his mixture control to get [the] best engine operation at altitude, resetting his instrument panel's gyro compass and all else vital to making the flight smooth and safe was, with Cornish, absolute clockwork.[34]

Yet despite Cornish's impeccable skills and his focus on safety, he was not immune from mishaps. He lost two airplanes, one of which was an OX-5-powered Waco that he was demonstrating for the Burgoyne Hardware Company in Hicksville, Ohio, on August 9, 1928. After failing to recover from a stall initiated too close to the ground, he crashed into a wheat field. The Waco "really folded up" when it pancaked in, he said, sending him face-first into the panel. "You can rest assured that the customer, Glenn Burgoyne, also slightly injured, did not buy."[35] Cornish walked away from his crumpled aircraft, its wings broken and its propeller smashed, with only a gash across his face.

By the late 1920s and early 1930s, navigational instruments and improvements in fuel efficiency had begun to make cross-country flying less of an art and more of a science. On March 13, 1930, Cornish took off from Guy Means Airport and

Cornish survived this crash near Hicksville, Ohio, on August 9, 1928. ("Plane Crashes near Hicksville," *Journal-Gazette*, August 10, 1928, 10. Courtesy of the *Journal Gazette*, Fort Wayne, Ind.)

stayed aloft in a Challenger Robin for sixty minutes on five gallons of gas before the motor began sputtering at a height of one thousand feet. For the record, he did not count the time it took to glide to the airport, but he did have to turn on the reserve tank to taxi to the hangar. The *News-Sentinel* did a story on the flight, reporting that he had set an unofficial world efficiency record for stock planes powered with 170 horsepower motors.[36] The following month, on April 20, he flew the *News-Sentinel* publishers to New York to attend the annual convention of the American Newspaper Publishers' Association. The total air time was seven and a half hours, with stops in Cleveland, Buffalo, and Albany for fuel and up-to-date weather reports. Less than two weeks later, on May 1, he flew a Travel Air Cabin monoplane powered by a Wright Whirlwind engine from Fort Wayne to Detroit in forty-three minutes, "chased through the air with a tail wind of an estimated speed of 100 miles an hour."[37] The strong tailwind at two thousand feet, he reported, enabled him to average 202 miles per hour. The return trip was much slower and unusually rough. And on June 15, 1931, Cornish made what was believed to be a record flight from Fort Wayne to St. Louis, accomplished at minimal cost: $4.55. Flying in a Curtiss-Wright Junior training plane powered by a 35 horsepower Szekely engine, he covered 350 miles in four hours and forty minutes, using seventeen gallons of gasoline and one quart of oil.[38]

Cornish credited pilot Jimmy Doolittle with demonstrating the efficacy of many of the technological and navigational aids that were now commercially available. But there was also a lot of experimentation in those early years. Cornish and Magnavox cobbled together a device using a windshield wiper motor that enabled instrument flying on radio ranges. He flew "under the hood" down to McCook Field to demonstrate the device to some of the fellows he knew there—"Mateland and Heggenburger and those birds." As he proudly recounted years later, his radio contraption worked. "Those days we had a lot of fun. We had to experiment around and we did a lot of it."[39] Muncie industrialist and veteran pilot Edmund Ball included a sketch in his book *Rambling Recollections of Flying and Flyers* of a crude "homemade" artificial horizon indicator using a Ball jar half-filled with a viscous liquid and a plumb bob hanging on a string outside the jar. "With practice and concentration it worked," Ball claimed. It helped when "flying blind" in clouds and other obscuring conditions. Even the best of pilots would become disoriented in a matter of seconds when flying "by the seat of their pants," Ball wrote.[40]

Cornish's job at Aereco required him to work seven days a week (including holidays), and often nights as well. This demanding schedule could have placed a strain on his new marriage, but the young couple thrived, both personally and professionally. There were perks for them both, including the occasional opportunity to use a company aircraft for personal pleasure, which they put to good use on a weekend trip with friends from Warsaw, Indiana, to Detroit, Ontario, Canada, and Buffalo. Clarence and Lois also took a midwinter vacation to Mount Dora, Florida, in 1930 to visit his parents, Frank and Ada, who had moved down there in 1926 to assist Irma's husband, Martin Hockemeyer, who worked as a contractor for the Lakeside Inn. The closest airport to Mount Dora was in Mount Plymouth. On January 18, Clarence and Lois flew in at near treetop level under a layer of low clouds and landed on the strip, which was scheduled to be dedicated that day at the new Mount Plymouth Aviation Country Club. Because of foul weather, the Curtiss Robin plane was the only one to land, but the celebration went on as planned, and all four Cornishes were guests at the party that evening.[41]

A year and a half after the founding of Aereco, Wall Street crashed. Initially the economic downturn did not seem to put a dent in flight activity in the Summit City. A January 17, 1931, headline in the *News-Sentinel* proclaimed: "Great Growth in Aviation Shown Here during 1930: More than Ten Thousand Passengers Carried and More than Three Thousand Miles Flown by Pilots at Fort Wayne Airports during the Year." Aereco was responsible for 57 percent of those passengers, having carried 5,896 individuals; Wayne, the other flying service, carried 2,900; privately owned Yankee Clipper carried 1,529. The

Lois and Clarence stopped for an overnight visit in Birmingham, Alabama, en route home from Florida. They stand arm in arm beside Aereco's Curtiss Robin aircraft on February 1, 1930. (Author's collection)

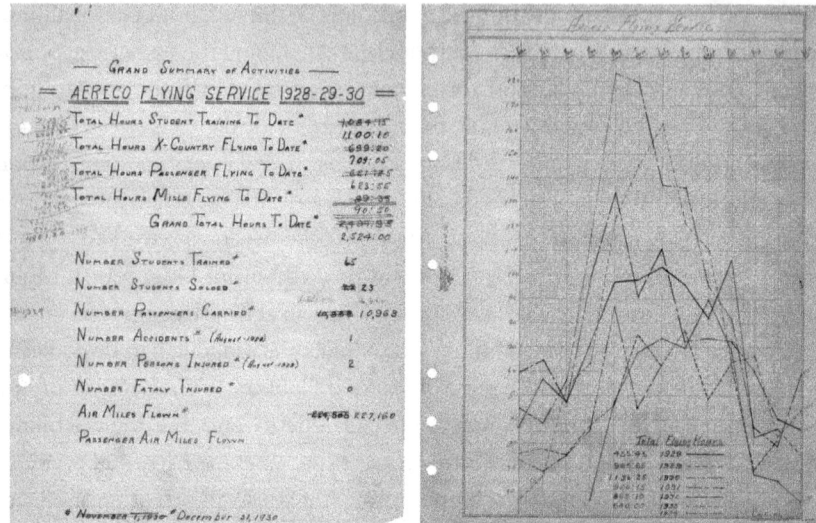

Cornish kept careful records, such as these from Aereco for 1928–1930: "Number accidents, 1; Number Persons Injured (August-1928), 2; Number fatal[l]y injured, 0." (Author's collection)

newspaper article listed seventy-one students and pilots, including Cornish, who were active in transport, limited commercial, and private flying. The city had flourished during the 1920s, during which time it expanded its municipal facilities (new bus service, municipal airport, new roads, better sewers) and attracted eight new industries, the largest of which was the International Harvester Corporation. Out of a population of 115,000 in April 1930, only 5 percent of the workforce was unemployed. Magnavox built a factory that year and added one thousand new jobs.

However, Fort Wayne would not be spared from the Great Depression. As community leaders became increasingly preoccupied with the problem of the unemployed, they wanted no help from the federal government. Said civic leader Arthur F. Hall, "No-one wants Governmental Taxation or Governmental Supervision for the care of the unfortunate."[42] In keeping with the "we can handle it ourselves" spirit, the city sponsored air shows for relief of the needy. In 1932 relief funds for food and clothing "swelled" by $500 through the efforts of Fort Wayne aviators and reserve Army airmen from Schoen Field in Indianapolis. The organizers turned over the profits to the Federated Relief Agencies to reduce suffering among the city's less fortunate inhabitants.[43] The next year's Sunday event, on October 22, raised significantly less, perhaps an indication of things to come.[44] By 1933 the city was in crisis mode because of the plight of banks and mortgages. Local resources were inadequate to deal with the situation, and Fort Wayne was forced to turn to Franklin D. Roosevelt's New Deal for help. Life would only gradually improve.

Cornish kept hand-drawn graphs that documented the hours he flew during his nearly seven years of management of Aereco. By every one of his measures, flying activity peaked in 1930; then the Depression began to take a toll on personal and commercial discretionary income. It certainly had an impact on Lois and Clarence, as he recalled:

> I was with Aereco Flying Service. And believe me, what people didn't need were flying lessons and buying airplanes during that period. I was lucky if I could take out enough money from the business to buy groceries. And I finally got to the point where I had to go to the folks at the Lincoln National Life Insurance Company who had the mortgage on our house, fully prepared to turn the keys to the place over to them.[45] We just couldn't make our payments. I knew the people there and the man that I was talking to, when I explained my situation, said, "Well, Cornish, we don't want your house. We're going to declare a moratorium on your house payments. And when you get on your feet again, you start making your payments again." Well, I can tell you that a few tears rolled down my cheeks because I was fully prepared to render the house to them and move out.[46]

Lois was the one who counted the pennies in their family, and her recollection differed:

> Well, it goes back to what I think I mentioned another time. When you're young, these things are not as important to you as when you are older. Although we didn't buy anything that we didn't need, I can remember one time when we thought we were really splurging and had two beautiful pork chops. I remarked to Clarence, "Just think, these pork chops cost $.25 a piece!" But I wasn't too concerned. I didn't get any gray hair over it. Maybe it was because I wasn't out earning the living. But we got through it because we didn't have debts [other than the house]. This is the whole thing.
>
> And then, of course, it was very shortly after that Mother and I visited Aunt Fanny, my grandmother's youngest sister [in Lansing, Michigan].

Before we were ready to come home, she and my Uncle Frank gave me a check for $500 and also the same amount for [my sister] Ruth. I immediately brought it right home and we scratched off the months—months, months, months of payments for our house. I think all that money went to reduce the mortgage of our house. . . . I really shed tears when they handed me that check and it was a most generous gift. At that time it was wonderful.[47]

Even as Wall Street crashed, a Ryan Brougham B-5 plane with a Wright J-5 engine arrived in Fort Wayne in the fall of 1929. It had been purchased from the Mahoney-Ryan Aircraft Corporation in St. Louis for $9,700 by community leader Oscar Foellinger.[48] In 1921 Foellinger had become the owner and publisher of the city's *News-Sentinel*, a paper that reflected his conservative views. An ardent Republican and a friend of Herbert Hoover, he managed the latter's Indiana presidential campaign in 1928.

Oscar, his brother Adolph, and Cornish were passengers on the aircraft's maiden voyage. The chief test pilot for Mahoney-Ryan, Captain Russell Young, flew them from Fort Wayne to Champaign, Illinois, where Oscar's daughter Helene was studying mathematics as an undergraduate at the University of Illinois. The high-winged, strut-braced monoplane, with a fully enclosed cabin for the pilot and four passengers, was christened the *Yankee Clipper* on November 25, 1929, when Helene, home from school, broke a bottle over the propeller.

Oscar Foellinger stands by his Ryan B-5. (*News-Sentinel*, November 11, 1929. Used with permission from the [Fort Wayne, Ind.] *News-Sentinel*.)

Close to five thousand people, including Lois, braved subfreezing temperatures and blustery winds to attend the dedication at Guy Means Airport.[49] Foellinger contracted with Aereco to service the plane and with Cornish to be his chief pilot. "Many memorable trips in this airplane were made," Cornish wrote much later when recording his life's story, "including several with Mr. Foellinger to Washington, D.C., to visit President Hoover and members of Congress. It is interesting to note that I was always Mr. Foellinger's guest and companion and not just his pilot on these trips."

Cornish went on to talk about another of Foellinger's acquisitions. "The newspaper also purchased a German-built primary glider which we [Aereco]

Every airport has its history, including the facility in Arlington, Virginia, that preceded Washington National Airport. Two fields built in the mid-1920s were combined to create Washington-Hoover Airport. When Cornish first flew to Washington as a private citizen in the 1930s and early 1940s, he used Washington-Hoover's 2,400-foot sod runway, which was bisected by the busy vehicular Military Road. The system was hardly foolproof, and reportedly the sheriff was occasionally dispatched to "shoo" cars off the runway. Cornish's perspective as a pilot was, "A green light meant you could take off; a red light meant you had to wait for automobile traffic to pass. That was air traffic control." As David Brinkley writes, the airport "consisted of little more than a grassy field between an amusement park on one side and a warehouse and dump area on the other. . . . The airport authorities, fearful of a collision between a car and an airplane, once asked the Arlington County government to close the highway for safety reasons. The county refused. The airport manager, on his own, installed a red light to stop automobile traffic when planes were landing and taking off. He was hauled into court, charged with installing a traffic signal without authority, and found guilty. The light was removed. For years, pilots flew in and out of Washington timing their landings to coincide with breaks in the traffic on Military Road, while using Hoover Field's only navigational aid, a wind sock nailed to a pole on top of the amusement park's roller coaster."[50]

When Hoover and Washington Airports merged in 1933, the nation's capital was known for having "the poorest aviation ground facilities of any important city in the United States or Europe," but despite its sorry reputation, the airport was not razed until after the completion of Washington National Airport on July 16, 1941, and after construction had begun on the Pentagon. Even though a notice was posted on September 23 of that year that flights into and out of Washington-Hoover were terminated, some pilots ignored the order, tearing up the runways with their heavy equipment.[51]

operated at Guy Means airport for a year or so. The glider was later given to Purdue University for use in organizing the Purdue Glider Club." The idea of forming a Glider Club in Fort Wayne had gelled after Cornish flew Foellinger and others to the International Aircraft Exposition in St. Louis, Missouri, held February 15–23, 1930. While they were there, gliders were one of the things that the men discussed with aviators and aeronautical experts.[52] The *News-Sentinel*'s glider was christened *Lindiana*, in tribute to Charles and Ann Lindbergh as well as the state of Indiana; that had been the second choice for naming the *Yankee Clipper*.[53] *Lindiana* would be "the first motorless aircraft in this part of the country," and John Emerson Pratt, considered to be Detroit Aircraft Corporation's most competent expert in glider training, spent several days in Fort Wayne training Aereco's pilots in May 1930.[54] Fort Wayne's Sky Hawk Glider Club held the state's first glider meet on September 20, 1931, at Paul Baer Municipal Airport.[55]

Lindiana captured the public's attention in the 1930s, but it was the *Yankee Clipper* that enjoyed greater and ongoing significance. The *News-Sentinel*'s Ryan B-5 was flown not only for commercial and political purposes, but to build community spirit; in its first year, the *Yankee Clipper* flew 221 hours and carried 1,472 passengers.[56] Foellinger wanted to provide unique experiences for disadvantaged children and to celebrate young people's achievements. In the latter months of 1929, he began to provide rides in his airplane for a group of children from the Indiana Masonic Orphanage in Franklin; dozens of honor students and their principals from Central, Northside, Southside, and Central Catholic High Schools; and members of Central High School's basketball team, the Tigers, winners of the regional title.[57]

Margaret L. Brudi, a senior at Northside High School, rode with Foellinger on Sunday afternoon, December 8, 1929, after which she wrote to thank him for making the flight possible and shared her impressions—the thrill of leaving the ground, the sensation of feeling suspended, the toylike appearance of houses and buildings, the "machines" along the highways that resembled crawling bugs, plunging into misty banks of clouds and flying blind, and, finally, her reluctance to return to earth. She had absolute confidence in the pilot, Mr. Cornish, she added, "and had no fear whatever."[58]

Following the first of the many flights that Foellinger sponsored and Cornish piloted, the *News-Sentinel* reported:

> Up a thousand feet was a solid cloud bank which extended upwards for another thousand feet. The passengers started out from the ground

with only a dreary gaze meeting their eyes. Plunging through the cloud banks, the *Yankee Clipper* carried its sight-seers to an altitude of about three thousand feet, one thousand feet above the clouds. Here the sun was shining brightly on the clouds below. It was the first time for most of the air passengers that they ever witnessed such a sight or even one equal to it in beauty.[59]

As part of Foellinger's commitment to rewarding outstanding effort and dedication, he asked Cornish to fly four "high-point" news carriers, Carl E. Doherty, Warden Lassen, Willard Lenz, and John Littlejohn, to Dayton and Wright Field, the largest experimental airport in the country and operated by the U.S. Army Air Corps. Four other planes joined the excursion, while several individuals drove over in their cars. Thirty in all enjoyed a guided tour. Since the city of Troy was only twenty miles north of Dayton, Cornish added a stop there to the day's itinerary so that the news carriers could tour the Waco Aircraft Company and see how various parts were made and assembled. On the way back to Fort Wayne, flying south of St. Mary's Lake, Cornish and his four passengers witnessed a scene that had seldom been viewed from the air. Circling around a plume of rising smoke, they all had a good look out the plane's windows at a barn engulfed in flames. They could see that the situation was unfortunately "beyond control of the farmers who were standing by helpless with buckets."[60]

In the spring of 1931, Oscar Foellinger proposed a grand and unprecedented trip that would take him and other Fort Wayne notables to the West Coast in the *Yankee Clipper*. With Cornish as pilot, Foellinger, Robert Klaehn (the owner of Klaehn Funeral Home), and Beach Hall (a local auto dealer) set off from Guy Means Airport on the combined business and pleasure trip a little after 6:00 a.m. on March 18; they expected to arrive in Los Angeles around 3:30 p.m. the next day. Foellinger wired the *News-Sentinel* frequently with updates. They had flown through driving snow between Wabash and Lafayette and in rain over Illinois, but the ship had "performed beautifully; everybody happy." After stops for fuel and meals in St. Louis and Tulsa, the men made it to Amarillo, where they spent the first night. (Flight time that first day was "exactly" ten hours and twenty-three minutes.) On day two, severe sandstorms kept them grounded in Albuquerque, where they had stopped for fuel and weather data. On day three, the men intended to fly over the Grand Canyon so that Foellinger could take some photos for his newspaper's rotogravure section. That plan was thwarted when their aircraft experienced a structural failure.

The *Yankee Clipper* flipped upside down when the left axle broke upon landing in Winslow, Arizona. (Author's collection)

On March 20, the *Yankee Clipper*'s left axle "gave way" upon landing at Winslow, Arizona, and the plane flipped upside down. Far from home, these prominent Fort Wayne men escaped with only scratches and one broken rib. In his meticulously documented log book, Cornish added a note in small script to his entry for that day: "axle broke—cracked in landing." "That serious consequences did not result was due to the quick action of Captain Cornish, cutting the switch," Foellinger wired home. Foellinger and Klaehn finished their trip west on the Transcontinental & Western Air route to Los Angeles; Cornish and Hall stayed in Winslow to oversee arrangements for the *Yankee Clipper*. The four returned home "with their enthusiasm for aviation undiminished."[61]

When the Kroger Grocery and Baking Company celebrated its fifty-second anniversary in 1934, it sponsored a series of events in twenty-two major U.S. cities. Cornish oversaw Fort Wayne's event, an air carnival held on Sunday, May 6, when twenty thousand observers thronged the airfield to watch exhibitions by more than two hundred planes, their pilots in "formal attire" (shirts and ties). Cornish, O. B. "Red" McVey, Lloyd Pierce, Eugene Campbell, and Harold Preston competed in a fifteen-mile closed-circuit race, with a three-foot-tall trophy to be awarded to the winner. They reportedly added thrills to the race by flying their "ships" close to the crowd on the straightaways and rounding the pylons in highly banked curves. The "victor" was Cornish in the *Yankee Clipper*.[62]

Air races such as the one in Fort Wayne were popular in the 1930s. That was especially true in Cleveland, Ohio, where some years as many as one hundred thousand spectators attended the National Air Races. There the coveted

Thompson Trophy was awarded annually to the winner of the ten-mile closed-circuit race, an event that featured aircraft competing at speeds over two hundred miles per hour around fixed fifty-foot pylons—the pilots fighting for fame and fortune.

Cornish's golden years as competitor, instructor, and chief pilot for the *News-Sentinel* ended soon after he assumed the managerial duties at Paul Baer Municipal Airport. His final pilot's log entry for the Ryan B-5 is dated September 24, 1934. Oscar Foellinger, the man whose enthusiasm for flight had inspired him to bring the plane named *Yankee Clipper* to his beloved city and, in the spirit of community goodwill, to share it with the community, died two years later; Helene, his daughter, then became publisher of the *News-Sentinel*.[63]

During most of the *Yankee Clipper*'s active years, it and Fort Wayne's aviation interests generally benefited from nationally mandated navigational aids. Airmail was the driving force behind this movement, and by the 1930s it was becoming as important as high-speed Internet service would prove to be at the turn of the twenty-first century. Round-the-clock transcontinental airmail service was begun on July 1, 1924, with a mandated system of lights and flares on aircraft and the placement of lighted airway beacons along sections of a main route, enabling pilots to see their way at any time of the day or night. The Air Commerce Act of 1926 gave an additional boost to air transport when it authorized the secretary of commerce to establish airways for mail and passenger service, to organize air navigation, and to license both pilots and planes. (The system of lights, or beacons, eventually expanded to eighteen thousand miles of airways with between fifteen hundred and two thousand beacons by 1933.) In January 1929, the transcontinental lighted airway from San Francisco to New York City was completed, using a system of high-intensity rotating beacons to guide commercial pilots flying at night from one city airport to another. Given good visibility, at an elevation of three thousand feet, the pilot could easily see the next beacon with its thirty-six-inch clear lenses.[64] For the early passengers, "flashing light beacons served as a visual reminder that the plane was on course—a reassuring thought as the aircraft winged its way through the inky black of night."[65] Other beacons ensured safety in the skies at and around airports, while landmark beacons warned pilots about navigational hazards. This lighted system was dependent upon good visibility.

Fort Wayne was ready to begin flying the mail, and its inaugural dedication of airmail service was scheduled for Saturday afternoon, December 6, 1930. Paul Baer Municipal Airport had undergone extensive upgrades to meet the federal requirements: new runways and concrete taxi strips were added, a new hangar was built, and a weather service was established that would become known as one of the finest in the Midwest. Paul Guild was chosen to head the ceremony planned for the inaugural event, with Cornish chairing the local pilots' committee. Other committee members were Harold Preston, chief pilot of the Wayne Flying Service; Warner Ashley, pilot of the Guy Means Airport; and O. B. McVey, pilot of the Paul Baer Airport. The plans included having local and visiting airplanes fly in formation over the city and having James H. Doolittle and James G. Haizlip, manager and assistant manager, respectively, of the Aviation Department of Shell Oil Company in St. Louis, buzz the field with various stunts.[66]

Greenville, South Carolina, had sent seventy-five thousand letters on the first airmail flight from that city. Not wanting to be outdone, Fort Wayne set a goal to surpass that number. Businesses were encouraged to hold back as many of their billings and advertisements as possible until the big day. By one account, schoolchildren even wrote letters to their grandchildren fifty years hence, presumably addressing the envelopes to themselves as historic keepsakes for their future descendants. In the days leading up to December 6, officials received ten thousand letters from all parts of the United States and Canada to be carried on the first airmail plane. Ultimately, however, the effort fell short, and only forty-seven thousand pieces were bagged to go out.[68]

Ben F. Myers, assistant superintendent of the U.S. Airmail Service, who served the entire Midwest and oversaw all airmail operators flying out of Chicago, was an honored guest at the event.[69] The plane designated to carry

> Montana's airway beacons are the last survivors of a great system of more than fifteen hundred that once dotted the country. Initiated by the Postal Service, which realized that airmail could offer no speed advantage if airplanes were idle during the night, the beacon system grew from a 1919 experiment with strings of bonfires to guide airmail pilots across the Great Plains into an eighteen-thousand-mile network of federal airways managed, after 1926, by the Department of Commerce. It survived into the 1970s, though by then few pilots were aware of it. When the Federal Aviation Administration decommissioned the beacons, Montana, which had thirty-nine of them marking routes through the mountains, took over those within its borders. Ultimately, it kept seventeen operating in its mountainous western half, linking Coeur d'Alene, Idaho, and Missoula, Helena, Great Falls, and Butte, Montana.[67]

the mail that historic day arrived from Columbus, Ohio, around 3:30 p.m., with departure to South Bend (and then by train to Chicago) scheduled for 5:22 p.m. Sadly, the weather failed to cooperate, and officials were forced to cancel all stunts and flyovers. Zero visibility kept the mail on the ground both Saturday and Sunday. Cornish and Eugene Campbell of Aereco did their best on Sunday afternoon to entertain the visitors by making a few rounds over the municipal airport, but they were forced down when officials had to cancel all the stunts and flyovers because of the worsening conditions.[70] "To everyone's humiliation, that night Fort Wayne's first 'airmail' went out on the Pennsylvania Railroad."[71] The weather remained inclement for the next two days, but nevertheless, and without fanfare, Lieutenant George Hill managed to fly part of the mail to South Bend late the following Monday. By the 8th the weather had cleared, and several round trips by air transported what remained of the original 47,000 pieces of mail to South Bend.

According to Roger Myers, author and curator of the Greater Fort Wayne Aviation Museum, Hill "made two or three round trips a day to South Bend" until April 21, 1932, when he became the first person to die in an aviation accident at the airport. "Coming back from his flight to South Bend, for some unexplained reason, it was clear weather and everything else, he just nosedived into the terrain and he was instantly killed." Added Myers, "His remains were interred in Cleveland. He left a wife and two children." In 1980, to commemorate the fiftieth anniversary of that first airmail flight out of Fort Wayne, Myers carried a pouch of mail to South Bend from and to the same airports in an open cockpit.[72]

For numerous reasons, most related to complaints of fraud and collusion on the part of commercial airmail carriers, private airmail service ended on February 9, 1934, when the government canceled all airmail contracts. Ten days later, the U.S. Army Air Corps took over airmail operations nationwide. It was a flawed decision that would lead to disastrous results. The Army pilots had no experience in flying the mail, and their planes were not properly equipped. Moreover, they were forced to assume responsibility for the unfamiliar air routes during an especially harsh stretch of winter weather. As a result, there were numerous accidents, including some fatalities. In a speech to the Hi-Y Clubs in March, Cornish explained why it had been a mistake to replace commercial pilots with Army pilots.

> In the first place, commercial pilots will do their best to get the mail through, but when they are licked, they know it and don't hesitate to

"bail out" and let the mail take care of itself. Army pilots had orders to protect the mail at all costs, and they obeyed those orders. Commercial pilots feel that their lives are of more importance than a penny postcard. Then, too, the commercial pilots have had years of training with constant flying over these same courses, and their planes are equipped with radio to guide them on their courses. Army pilots, prior to the order to fly the mail, were allowed only 10 hours of flying a month—120 hours a year. Their ships were not equipped with radio, and they had to rely on their compasses and flying instinct to keep to the course. By means of their radios, commercial pilots are in constant contact with the ground and are provided with "beams" to guide them. These young fliers have been given plenty of training in war flying, and although I have had many years of flying experience myself, I would not go up to fight one of them. It would be suicide. But they have not been trained in long-distance flying.[73]

By May, the federal government had reversed its earlier decision. While Congress wrestled with a longer-term solution to the problems with commercial airmail, Postmaster General James A. Farley was permitted to award temporary contracts to private carriers. In Fort Wayne, politicians, community leaders, businessmen, and aviators began working together to bring back local airmail service. Mayor William J. Hosey wrote, "Our airport is not merely a makeshift landing field—it is a complete and well-planned airport, lighted and maintained for all-weather flying. This ought to be taken into consideration by the postal authorities when they consider making new airmail contracts." Cornish worked actively to establish an airmail line through Fort Wayne. On May 23, 1934, he flew the Ryan B-5 to Washington with Paul Guild, chair of the

> Between 1928 and 1934, Cornish flew the following aircraft:
> With Aereco:
>   Waco-10 with OX-5 engine
>   Waco-10 with R. Siemens engine
>   Waco-10 with Siemens-Halske engine
>   Curtiss Robin with OX-5 engine
>   Fleet with Kinner K-5 engine
>   Travel Air with OX-5 engine
>   Curtiss Robin with Challenger engine
>   Travel Air with J-6-240 engine
> With the *News-Sentinel*:
>   Ryan B-5 with J-6 engine
>   Ryan B-5 with J-6-300 engine

Fort Wayne Chamber of Commerce Aviation Committee, and senatorial candidate R. Earl Peters to meet with Farley and plead Fort Wayne's case, asking to have the city added to a Detroit, Toledo, and Indianapolis airmail route. However, the Post Office Department soon rejected the request.[74] It was inevitable that airlines would carry airmail, but it would not happen immediately in Fort Wayne.

The Air Mail Act of June 12, 1934, once again allowed private carriers to bid for new contracts, but airlines no longer viewed carrying the mail as a sufficiently profitable activity. They were more interested in expanding passenger service. The rapid growth of passenger airlines through the mid to late 1930s would place new demands and expectations on airports. At this critical juncture, the Fort Wayne Board of Aviation Commissioners appointed Cornish to manage the city's airport. Leaving behind Aereco, the aviation service he had nurtured from its inception, he assumed his new position at Paul Baer Municipal Airport, ready for the new challenges ahead.[75]

5

Fellowship Forged through Flight

By the late 1920s, the lure of flying was drawing men like bees to pollen.¹ Although flying and related activities were growing rapidly, the fraternity of fellow flyers was small enough that it was easy to know everyone who was involved at the local level. Even as aviation blossomed, most of the participants at the state level were still well known to each other. Cornish was primary among the many who built associations beyond their own communities; he would one day rise to the top. Those connected with flying, from mechanics to pilots to builders to sellers, sought one another's company in their free time and enjoyed the fellowship that develops easily among like-minded people. Newly founded groups such as the Fort Wayne Aero Club and the Indiana Aviation Trades Association brought together those who were enamored with flight.

On September 27, 1933, the Fort Wayne Aviation Club held a party "to give the club members the benefit of the amount left in the treasury of the club" before dissolving after several years of activity.² It was replaced that night by a new club. The members initially referred to the new entity as the Aereco Aero Club, but it was almost immediately renamed the Fort Wayne Aero Club.³ It became part of a loosely connected worldwide group of clubs with varying individual missions. Bob Schott was the first president of the local group, and Cornish was an active participant in this assemblage of local flying enthusiasts who dedicated themselves to promoting and creating interest and safety in aviation—and having fun.

The Aero Club maintained a full social calendar for the aviators, students, and spouses who were connected with it. On July 28, 1934, during the club's first summer, the Chicago Girls' Flying Club gave a demonstration, which was followed by a dance in their honor at the municipal airport.[4] The next month's activities included a golf game. Thirty of the club's pilots took off one at a time, flew over two golf greens, dropped golf balls enclosed in bags attached to parachutes, then landed and played the balls "in the usual manner, using a regular putter." Amby Babbitt won the title of "City Champion" with eight shots.[5] Cornish, who would become an avid golfer in later life, was away that weekend, practicing training maneuvers with the U.S. Army Air Corps.

The following short list is typical of the social gatherings organized by the Aero Club during the years leading up to World War II: a cabaret-style dance at Valencia Gardens (part of the Scottish Rite Auditorium); a roller-skating party at Bell's Rink that netted the club $18.10; a smelt fry; a flying cruise to D. D. "Doc" and Mrs. Johnston's cottage on Big Long Lake when the ice was thick enough for landing; and a hunting trip that concluded with a wild game dinner and dance at the Moose Lodge. One fall, the club members headed back to Johnstons' cottage for a wiener bake, followed by dancing to the music of a three-piece band at Lake of the Woods. These social occasions provided light relief from the more serious purpose of the Fort Wayne Aero Club, which organized goodwill northern Indiana air cruises, informational programs, and philanthropic events, including the annual air circus, held the third Sunday in July.

The first air cruise took place on October 12, 1935, when ten planes carrying twenty-five pilots and passengers formed the largest armada of local aircraft that had ever congregated for a mass flight. A news article reported, "The first ship to take off was a Stinson, piloted by Capt. Clarence Cornish, cruise director, who is blazing the trail of the cruise." The 1936 cruise, on August 22, "got under way at 9:15 o'clock when Capt. Clarence Cornish, airport manager and cruise director, roared down the run way in a U.S. Army observation plane from Schoen Field, Indianapolis, and pointed the nose of his ship towards Goshen." Both years the *Journal-Gazette* awarded safety trophies, which were displayed beforehand in the window of the Baber Jewelry store; Herbert V. Koch was the first winner, and Don White the second. Awards were given to the five pilots who the judges felt displayed "the best flying skill, judgment and sportsmanship on the cruise."[6]

Cornish was an events announcer for the 1935 air circus. It attracted an "orderly throng" of fifty thousand spectators (twenty-five thousand paid admission onto the grounds of the municipal airport), who were reported to have waited until the final event before surging "on[to] the field to get a glimpse of some of

A crowd of some 50,000 enjoyed the Fort Wayne Aero Club's 1935 air circus. Air events in the 1920s and 1930s often attracted tens of thousands of spectators. ("Thousands Thrilled at Air Show Here Sunday," *News-Sentinel*, July 15, 1935, pt. 2. Used with permission from the [Fort Wayne, Ind.] *News-Sentinel*.)

the fastest speed ships in competition." The event netted $4,000, divvied up as follows: 5 percent to the Aero Club, 5 percent to the Board of Aviation Commissioners, 10 percent to American Legion Post No. 47, and the remainder to the show's organizers, Detroit's Skywriters Inc. The latter amount, $3,200, was shared among those who took part, described as the country's "most famous speed and stunt pilots."[7]

The Aero Club's portion helped pay for educational entertainment, which was free and open to the public. One program offered a showing of *Flying the Lindbergh Trail*, an "aerial travelogue" of South America that included scenes from more than thirty-five countries and colonies of the West Indies and Central and South America and gave viewers their first opportunity to see "many subjects which, before the coming of the airplane, have been accessible to few white men." Another program featured photographs of Cleveland's National

Air Races and of Roscoe Turner, the winner of the Thompson Trophy for long-distance speed, in his Hornet-powered Wedell-Williams.[8]

The *News-Sentinel* gave the Aero Club excellent coverage leading up to the 1939 air circus, which was held on Sunday, July 16.[9] Upwards of 20,000 people thronged to the airport to watch the aerial feats. Bill Sweet Jr. was the announcer. Air show favorite Mike Murphy performed new stunts and flew at night in the *Marathon Flying Display*, with the edges of its wings outlined in lights, and the undersurface flashing messages in neon. Unfortunately, while the plane was parked and unattended that Sunday, it caught fire and went up in flames. The suspected cause was an improperly discarded cigarette.[10]

The same newspaper may have had that incident in mind when it reported early the following year, "Because of increased facilities and activities at the airport and the increased amount of property and number of planes, the present apparatus has been deemed inadequate for combatting fires. [With modern equipment] airport personnel on hand could rapidly respond to and safely haul fire-fighting equipment to the scene of aircraft-related accidents should they occur." By that time the Aero Club's treasury had grown to $1,300, and the members had already voted to donate $1,000 to the airport for the purchase of a modern, "all metal" Chevrolet Carryall fire truck.[11] "Such a substantial and significant donation," they felt, "would certainly establish this club as a civic-minded group and would gain prestige with the City Fathers."[12] It seems likely that the organization had already earned the city's gratitude, given the many public events it had initiated and held over the years.

Whereas the Fort Wayne Aero Club was open to any flying enthusiast, the Indiana Aircraft Trades Association (IATA) was a group for aviation professionals. Many years before the Aero Club was formed in 1933, Cornish and other Indiana flyers felt the need for a statewide organization. "After entering the business of a fixed base operator, Aereco, in the spring of 1928," Cornish later recalled, "it became obvious to me that such businesses should be organized on a State and National basis for their mutual protection and development. This belief led to the organization of the IATA that year."[13] Renowned aviator Harvey Weir Cook served as the first president. The organization would have a strong legislative component, in which Cornish played a vital role, but initially it strove to promote and showcase new aircraft and to help spread the word about aviation's commercial viability. IATA members believed that one way to accomplish their objective was to organize flying events that would stop at numerous Hoosier cities with reasonably good landing sites. Their instincts proved to be correct. For more than a dozen years, the All-Indiana Air Tour "brought together many of the state's leading aviation personalities" and helped build a camaraderie that would last throughout the lives of many.[14]

If asked to recall their fondest memories of their early flying days, the elite fraternity of those who were privileged to organize and participate in the air tours would undoubtedly hark back to those heady days. Though the events were staid in comparison to the daredevil air spectacles of the same period, appreciative crowds of local officials and the public at large came out to greet the flyers and their often notable passengers. Philanthropic and aviation-related organizations and political leaders in communities throughout the Hoosier state rolled out the red carpet for the pilots and their passengers when they landed.

The IATA's first tour was held in 1929. One of the tasks of W. F. Sturm, who served as director that year and would be flying as the passenger of Captain Charles E. Cox, was to set the stage for the coming attraction.[15] In September 1929 he lauded the state of aviation, remarking that "it had settled down to a business basis" and calling attention to "how the commercial pilot of today, an employe of a permanent flying field or of one of the great trans-country transportation units, handles his ship." Pilots, he continued, were past the "hairbrained stage" when they had found it amusing to frighten their passengers; they were now refocused on the primary objective of flying: to move people "from one point to another with the least possible maneuvering or anything that approaches a thrill for the passenger." He declared that "engines today are as reliable in the air as in automobiles—they don't stop suddenly. Air passengers can fly without the least trepidation. If a citizen of 1929 wishes to travel from South Bend to Kokomo, the trip could be made far above dust and traffic at a safe speed of 80 to 100 miles per hour." "So long as the plane is on the ground," he concluded, "it must have a hard surface to prevent sinking in as the car does. Otherwise, it cannot get up enough speed to get into its natural element—the air."[16] This was a truism that would be experienced later by the tour participants.

Fort Wayne's civic leaders gave the event their full support. The city was to be the eighth stop on the six-day tour, which would begin in Indianapolis, then move on to New Castle, Connersville, Union City, Richmond, Muncie, Anderson, and Kokomo before arriving in Fort Wayne. From there it would continue on to Peru, Plymouth, Goshen, Elkhart, Gary, Lafayette, Greencastle, Terre Haute, Princeton, Boonville, Evansville, and Bedford, before ending back in Indianapolis.[17] Leading up to this highly touted event, an article in the *News-Sentinel* included photos of the twelve "prominent men in aviation" who had been instrumental in bringing it to Fort Wayne: Cornish, Guy S. Means, Arthur F. Hall, Matt Jones, Ross McCulloch, Eric L. Hardy, Ben F. Geyer, Paul Guild, George DeWald, J. Ray Schromp, Chester G. Schiefer, and J. B. Wiles. These men, plus S. F. Bowser and Company and Robert E. Feustel, had helped to defray the local expenses.[18]

At 4:35 on the afternoon of September 17, thirty planes began a gradual descent, one after another, to the Guy S. Means Airport. To familiarize the public with the participants, the paper provided a full listing of the aircraft, pilots, affiliations, and, in most cases, passengers. The pilots were Cornish flying a Curtiss Robin with an OX-5 motor; Cox in a Curtiss Robin cabin plane with a 170 horsepower Challenger motor; R. S. Lamont in Standard Oil Company of Indiana's mammoth new Stanolind III with three 425 horsepower Wasp nine-cylinder radial motors (it must have been an attention-grabber at the various stops); Earl F. Ward in a Fairchild cabin plane with a 425 horsepower Wasp motor; Clyde Shockley in a Waco Taper-Wing with a Wright 300 horsepower motor; Robert Shank in a Travel Air biplane with a Wright 300 horsepower motor; Lieutenant Lawrence I. Arentz in a Curtiss Robin cabin plane with a 150 horsepower Hispano-Suiza motor; E. H. Jose in a Ryan Brougham cabin plane with a 200 horsepower motor; Freddie Lund in a Waco Taper-Wing with a 300 horsepower motor; Leland Jamieson in a Fleet Husky Jr. with a Warner Scarab 110 horsepower motor; J. H. Stewart in a Monocoach cabin plane with a 200 horsepower motor; Theodore Hubbell in a Monosport cabin plane with a Warner Scarab 110 horsepower motor; H. Weir Cook in a Curtiss Robin cabin plane with a 170 horsepower Challenger motor (his passengers were the mayors of each city or their representatives); Merlin Boyd in a Travel Air biplane with a Hispano-Suiza motor; French Livezey in a Travel Air biplane with a 90 horsepower motor; Paul Snick in a Swallow biplane with a Kinner 110 horsepower motor; D. K. Russell in a Waco biplane with a Hispano-Suiza 180 horsepower motor; D. Homer Stockert in a Fairchild cabin plane with a Wright 200 horsepower motor; George A. Retting in a Standard Swallow with an OX-5 90 horsepower motor; O. A. Brant in a Waco 90 with an OX-5 90 horsepower motor; Cy Younglove in a Davis V-3 with a LeBlond 60 horsepower motor; Walker Winslow in a Command-Aire with an OX-5 90 horsepower motor; Arthur F. Foulkes in an American Eagle with an OX-5 horsepower motor; Charles Bedell in a Travel Air with a Wright J-5 200 horsepower motor; Ted Lundberg in an unidentified aircraft; Paul Cox in a Barling monoplane with a LeBlond 60 horsepower motor; Walter McClain and Harry White in an Aristocrat cabin plane with a Warner Scarab 110 horsepower motor; Ray Kuhl in an

The air-marked roof of the Lincoln Life Insurance Company pointed the way toward Fort Wayne's municipal airport. (Author's collection)

Eaglerock biplane with a Hispano-Suiza 150 horsepower motor; E. W. Campbell in a Curtiss Robin cabin plane with an OX-5 90 horsepower motor; and Lieutenant Matt Carpenter in a Tour Master plane with a 200 horsepower Wright motor. The city feted the pilots and their passengers at a banquet later that evening at the Anthony Hotel. After an overnight stay, they left for Peru.[19]

Navigational aids for the average flyer were still rudimentary at the time of the inaugural tour in 1929. In keeping with the IATA's interest in boosting the ease and safety of flying for everyone involved, the organization encouraged Governor Harry E. Leslie to proclaim September 18–23 "Mark Your Towns Week," which he agreed to do. At the twenty stops along the tour route, pilots advocated for navigational markings in the form of town names painted on the roofs of prominent buildings, a relatively simple but important means of helping pilots stay on course when they were flying cross-country. The lettering was to be bold, ten to twenty feet high, painted in chrome yellow on black, the colors of the Aeronautics Branch of the Department of Commerce. "The moment a city is marked," Weir Cook said, "that moment the word begins to spread among the pilots who fly from place to place. Pilots, of course, do not depend on the names on the city's roofs to find their way, but it certainly does give them a warm glow about the heart when they see the progressive city sign and find they have hit it on the nose, which is the argot for going direct to the spot."[20] An inveterate promoter of flight, he emphasized that Indiana must remain at the forefront in aviation and that "aviation had come to stay as a means of rapid, dust-free, and safe transportation."[21]

One Indiana state official made the most of this first tour. Richard Lieber, director of the Indiana Department of Conservation, used it as an opportunity to have an assistant, Paul V. Brown, fly around the state and gather information. From the air, Brown would be able to scout out potential locations for landing fields in communities near some state parks and memorials. Lieber envisioned "enterprising air lines" that would operate out of Indianapolis and fly to cities near where state parks were located, thus encouraging "this form of transportation to these great Hoosier scenic preserves." During the flights, Brown also "observed thousands of acres of wooded lands contemplated of purchase by the department in its program of reforestation, establishing game preserves and building up an estate for the state."[22]

The Fort Wayne newspapers also reported on a "New Stowaway Stunt." An unnamed pilot had been assigned to fly Richmond's mayor, L. A. Handley, from Richmond to Muncie, the next stop on the tour. When a man climbed into the plane and began settling himself into the cockpit, the pilot, who was unfamiliar with the dignitary, mentioned that the seat was reserved for Mayor Handley. "I'm the Mayor," proclaimed the pretender. When the real mayor arrived, the deception was discovered, but by then it was too late: the plane had already departed. (Alternative arrangements were made.)[23]

The Indianapolis Chamber of Commerce was enthusiastic about the Air Tours, seeing them as a means to advance the economic development of major metropolitan areas around the state, including its own. "Your Chamber of Commerce has been the headquarters for all the preparations for the Indiana Air Tour, which is covering all of the leading airports of the state this month," the organization declared in June 1930. At the time, the opening of the Indianapolis Municipal Airport—which the Chamber was envisioning as "a great air center"—was only a few months away.[24]

In 1933, the air tour was directed by Herbert O. Fisher, who would go on to be a renowned test pilot and aviation executive. Cornish, after a three-year absence, returned to participate and carried two passengers with him. The tour began in Indianapolis on June 20. On the first day, well-known daredevil Gene Rock rode along, making parachute jumps at the noon and overnight stops. Fred W. Goebel, who represented the Berghoff Brewing Corporation, Cornish's sponsor, flew with him for the entire tour, including the stop in Fort Wayne on June 23.[25] The visit to Richmond that year was apparently memorable for more than the planes and the flyers. Reporters were particularly impressed with the lunch at the Richmond Airport, managed by Tat Lower and hosted by the Rotary Club and Merchants' Association. Mary Bostwick of the *Indianapolis Star* wrote that the meal was "served by such a bevy of beautiful damsels that the birdmen kept coming back again and again for refreshments." She added, "All the flying femmes were presented with gardenias."[26] Other "damsels" that year were women pilots Shirley McKittrick and Lenore Harper from Indianapolis and Gertrude Allen from Kokomo.[27]

In at least one way, Cornish found that year's tour to be exceptionally gratifying. Flying Aereco's Travel Air monoplane, he received what was regarded as the tour's highest achievement award, the All-State Air Tour Safety Trophy, which was presented each year by Fort Wayne's Lincoln Life Insurance Company. He and those so honored in other years, including Charles E. Cox, Walker Winslow, Robert F. Shank, Vic Lindemann, and Lawrence "Cap" Aretz, were chosen by a conference of judges for having shown the greatest regard for safety and having meticulously observed the rules throughout the tour. William B. F. Hall, the son of the insurance company's president (and a flyer himself), spoke of the qualities embodied by the award—"character, knowledge, application, ability, and sound judgment." He presented Cornish with the twenty-seven-inch-tall trophy, "a metal pillar surmounted by three silver eagles supporting a globe while the whole business is a nymph lovingly clasping an airplane in her arms."[28] Hall,

William B. F. Hall awards the 1933 All-Indiana Air Tour safety trophy to Cornish. Standing behind them are W. R. Newlin and Fred W. Goebel of the Berghoff Brewing Corporation. ("Cornish Home with Safety Trophy," *News-Sentinel*, June 27, 1933, 22. Used with permission from the [Fort Wayne, Ind.] *News-Sentinel*.)

his wife, and their police dog, Pittsburgh, were at the Hoosier Airport in Indianapolis when the award was made at the tour's conclusion on Sunday, June 26. Pittsburgh, wearing goggles for his flight, was also Aereco's mascot. A news photo shows him "sitting up for the pilots just before he climbed into the cockpit to take a ride with his master, Ensign William Hall, president [and owner] of the flying service."

In honor of Cornish and Goebel, Berghoff threw a party the next week, on Tuesday, June 27, at the company's plant, located at 1025 Grant Avenue. Tour officials Charles E. Cox, Herbert O. Fisher, Walker W. Winslow, Stanton Smith, Richard Coryell, Howard Maxwell, and Richard Arnett flew up from Indianapolis in a Fairchild monoplane. Forty civilian pilots from around the state joined the celebration, along with some army pilots, city and aviation officials, and the tour's committee members from Fort Wayne.[29]

"Pittsburgh" the new mascot of the Aereco Flying Service, Inc., is shown here sitting up for the pilots just before he climbed into the cockpit to take a ride with his master, Ensign William Hall, president of the flying service.

Aereco's mascot, Pittsburgh, donned goggles before climbing into an open cockpit for a flight with his owner, William B. F. Hall. (Unidentified newspaper photo)

The Air Tour Committee, of which Cornish was then a member, met in Indianapolis in February 1934 to begin making plans for that year's event. Launched September 10, the 1,000-mile journey included eighteen Indiana cities, with overnights in Seymour, French Lick, South Bend, Wawasee, Fort Wayne, and Muncie. September could generally be counted on to be a dry month in Indiana, but that September was not typical. Aircraft arriving at French Lick at the end of the second leg showed signs of having flown through heavy rain.[30] Some arrived directly from Seymour without making the scheduled stops at North Vernon and Bloomington. In addition, "after a country breakfast at the Graham farms near Washington, the giant tri-motored ship piloted by Ray Loomis of Indianapolis was forced to transfer two passengers to another ship when the large plane sank hub deep in the soft turf. Several others had difficulty getting off the ground toward their next stop, Lafayette."[31]

The tour continued on, and it was headline news when forty-five "ships" arrived at Fort Wayne on the 14th as "threats of rain and low ceilings and poor visibility hung over the city airport."[32] "Heading the galaxy of birdmen who landed in Fort Wayne were Lieut. Stanton T. Smith, commanding officer of Schoen Field, Fort Benjamin Harrison, and Herbert O. Fisher, director of aeronautics for the Indianapolis Chamber of Commerce." Others followed, most prominently Charles E. Cox, who was president of the IATA, superintendent of the Indianapolis Municipal Airport, and state Federal Emergency Relief Administration (FERA) airport works aide.[33] Cornish, flying a 300 horsepower Ryan cabin monoplane, was part of the mass descent at Baer Municipal Airport.

Performances for the crowd that gathered at Baer that day included a parachute jump by Gene Rock and a demonstration of formation flying by the Inter-City Flying Service. But most noteworthy were the hair-raising feats of Mike Murphy, manager of the Kokomo Airport, who "flew without hands, zooming over the tops of automobiles and fences with very little room to spare, while his hands hung inactive over the sides of the cockpit" and who then "performed a double loop and barrel roll with a dead engine, landing after a series of sensational side slips."[34]

The Aero Club was Fort Wayne's tour sponsor. The members welcomed pilots and passengers at a banquet at the Chamber of Commerce, with entertainment provided by magician Clifford Kirkpatrick and accordion soloist Aro Grunert. Special guest Jimmie Mattern, an internationally known long-distance pilot from Buffalo, New York, enjoyed the celebratory occasion, where he expressed his pleasure at being in a part of Indiana where people "do something other than call hogs and ride horseback."[35] The next morning, September 15, "amidst a popping and roaring of motors" and still challenged by inclement weather, the departing flyers, some with planes "still shiny in spite of a week of flying in the worst stretch of weather ever battled by pilots of an Indiana tour," departed toward Muncie and the final night of the tour.[36]

Plans for the seventh annual event began to take shape in late February when Cornish, with Paul Guild, the vice-president of the local Board of Aviation Commissioners, and Whitney A. Gregg, the manager of the Inter-City Flying Service, drove to Indianapolis to discuss plans. Due to sleet and icy road conditions, the men stayed overnight in the capital city. Cornish sent a wire to Lois on the 21st to tell her not to expect him back until the next day.[37] The 1935 event was marked by more "unfavorable weather," which "made cross country flying dangerous throughout most of the tour." Held June 17–23, the 1,500-mile tour began with forty-five planes.[38] Cornish, on his way to the opening-day festivities in a four-place Stinson owned by the Inter-City Flying Service, was temporarily delayed in Marion when he became socked in because of clouds and fog. As a result, he missed the ascent of planes, which took off in a steady stream from the Indianapolis Municipal Airport following an aerial bomb that sounded the 9 a.m. start.[39] He later explained that he did not want the student who was accompanying him in a separate plane to fly under such poor conditions; nor did he want to risk his own safety. They caught up with the tour, but heavy rains made cross-country flying and landing difficult, and in some cases impossible, that week.[40] At Winamac Airport, for example, the rain turned the field into a sea of mud, forcing the scheduled stop there to be canceled.[41] Given that most airfields were still sod-based and lacked paved landing strips, rain often hampered air events—although getting wet while in flight may not have been an issue for Mike Murphy, who had promised to fly the entire tour upside down in his taper-wing Waco Plane. (A later article stated that Murphy had indeed flown the majority of the route upside down.)[42]

Fighting cloudy weather and threats of rain in South Bend on the 21st, the tour made it to Fort Wayne that Friday afternoon. Cornish was the eighth to land, with passenger Ross McCulloch. Despite the fact that they spent only two and a half hours on the ground there, an article in the Saturday *Journal-Gazette* was headlined "4,000 Welcome State Air Tour at Baer Field." Once again, women had a noticeable presence. The caption for a photograph that appeared with the article touted the flawless piloting by Shirley McKittrick of Indianapolis and Alice Hirschman of Detroit. And the attendance of *Indianapolis Star* reporter Mary Bostwick was also noted; along with tour director Herbert O. Fisher, she had taken part in the event every year since its debut in 1929. In their view, "This year's aerial armada is the biggest and best yet, with the enthusiasm shown in the cities visited indicating gain in support and prestige."[43] Mayor Harry Baals credited the tours with having "drawn national attention to the progressive spirit of our state." The entourage left Fort Wayne for Lake Wawasee, where they spent the night as guests of the Spink-Wawasee Hotel. Although the weather conditions were challenging, the 1935 tour had been a success. At a meeting the following winter, the IATA announced that the eighth annual All-Indiana Air Tour would begin on June 22.[44]

A significant number of women, both pilots and passengers, participated in the 1935 All-Indiana Air Tour. (Author's collection)

One month before the 1936 tour began, the head of the IATA, Frank E. Ball of the Muncie Aviation Corporation, was killed when a faultily designed wing fitting failed on his new Waco F-2, causing it to crash and burn. A member of the prominent Ball family of glass manufacturers as well as a highly regarded Indiana pilot, Ball died instantly upon impact.[45] Even the most highly engineered aircraft could sometimes fail, with catastrophic results. On Monday night, June 22, the opening day of the tour, IATA officials paid their respects at Ball's tomb in Muncie.[46] He was one of three well-known Hoosier pilots who lost their lives in accidents within a year's time; the other two were Paul Cox of Terre Haute and Richard Arnett of Lafayette.

Despite the loss of those talented and dedicated aviators, the tour went on as planned—although things did not go as smoothly as expected for daredevil Mike Murphy and his passenger, National Air Races announcer Bill Sweet Jr. Murphy's plane developed a problem with the motor, forcing him to land it in a rhubarb patch near Indianapolis for repairs.[47] On the 27th, fifty planes arrived from Farmland at the Paul Baer Municipal Airport, where the city welcomed the "Air Armada." The Berghoff Brewery expected 150 pilots and passengers to travel by car to the Anthony Hotel, where they would be treated to a "Dutch lunch" and entertainment.[48]

Presumably the tour's mascots did not attend. Two animals flew in 1936—Lola Mae, a one-month-old registered Guernsey calf, "the principal attraction of the children at all landing spots of the tour"; and Linco, a tiny piglet only a few days old. Lola Mae, owned by the Guernsey Breeders' Association of Indiana, traveled on a cushion of hay, wrapped in large cloth sacks for warmth; she reportedly slept most of the time, "with her chin on the knee of her owner, M. W. Dunman, of Russiaville, who has been dubbed 'The Flying Milkman.'"[49] Little Linco was so named because Linco gasoline and oil was used exclusively by all of the tour planes. The Ohio Oil Company, producer of Linco products, hosted the tour in its hometown of Findlay, Ohio, on Friday, June 26.[50]

That year's tour was remarkable for another reason—there were three non-fatal crashes. The first accident happened near Hicksville, Ohio, when the motor of a Waco open-cockpit plane stalled at about three hundred feet and the aircraft landed on its side in a cornfield as the pilot attempted a dead-stick landing. The pilot and his passengers, all from Kokomo, suffered minor injuries. Three hours later, Clarence McElroy of Winamac, who was on his way to pick up the victims of the crash, hit a hole while landing in the field of the Hicksville Airport and nosed over, breaking a propeller.[51] Then on the 29th, at the Paul Baer Municipal Airport, the pilot of a Waco 10 and his passenger, both from Culver, "suffered multiple laceration[s] and abrasions on their heads and legs, when the motor of the plane in which they were attempting to take-off stalled just as it left the ground and crashed into a large tree about a thousand feet northwest of the landing field." Though the plane was severely damaged, the two men were treated at a Fort Wayne hospital and released.[52]

The early years of aviation in Indiana had been marked by a true sense of fellowship. In 1937, however, that camaraderie was suddenly shaken. Cornish and the IATA had taken the lead in advocating for the formation of a state aviation committee, but not all of their fellow Hoosier flyers were on board. A group of pilots who were opposed to the idea had therefore made the decision to form their own organization, the Indiana Air Pilots Association (IAPA). They now announced that they would hold their own tour in June, in direct competition with the traditional All-Indiana Air Tour. Alvin Tarkington of Indianapolis, who headed the group, said that they had been told that the IATA was giving up its sponsorship of the tour. "Incorrect," according to Cornish, who went on to suggest after an April meeting in South Bend that the two groups might cooperate and simply let the IAPA be the sponsor that year.[53] Somehow the issue was resolved. Ed Ball recounted that in 1937, fifty pilots (Cornish among them) and passengers attended the wrap-up banquet held in Muncie's Roberts Hotel on the evening of Saturday, June 19. He added that heavy fog off Lake Michigan had delayed the pilots' departure from the area. Consequently, "some landed in Muncie so late they came directly to the banquet from the airport still dressed in their wrinkled flying clothes and greasy coveralls."[54]

The 1938 tour took place later than usual in the year, beginning on October 10 and ending on the 14th; the number of days had been reduced from six to five, and there were fewer participants. This time the details about Fort Wayne's part appeared in the newspapers a mere five days before the event's launch in Indianapolis. Though thirty-five planes were expected in Fort Wayne the afternoon and evening of Wednesday the 12th, only twenty-five arrived. A Dutch lunch was served at the airport. Weir Cook's collection of scale model planes as well as first-flight envelopes, airmail literature, and other items were displayed at the Chamber of Commerce, available for the public to view that day only. The presence of Mary Bostwick was again worthy of note: she was flying in her tenth tour.[55] However, when the entourage landed in Muncie for the tour's last overnight stop, Ball recalled that only eighteen planes arrived.[56] The tour concluded in Indianapolis the next day with a banquet at the Hotel Washington, over which Cornish presided. The principal speaker was Oswald Ryan, a member of the Civil Aeronautics Authority in Washington, D.C., who discussed the government's new aviation organization.[57]

The All-Indiana Air Tour had peaked in 1936, when fifty pilots landed their planes in Fort Wayne and took part in the festivities that the city offered. They represented the largest number of participants in the event's history. During Cornish's tenure as IATA president, the tours continued, at least into 1940.[58] But there was no denying that interest had declined and local organizers had begun to lose their enthusiasm. In 1939 Fort Wayne sponsors grumbled about the tour's failure to follow through with its agreements. Even though the city was promised a mock air battle, parachute jumps, and autogiro demonstrations, scheduled for Friday, June 23, a group of men, including Cornish, were appointed to attend the planning conference in Indianapolis during the winter of 1939 to air their grievances and insist that the tour organizers follow through with their commitments. The *News-Sentinel* gave that year's tour no coverage, in sharp contrast to the treatment given to the Aero Club's air circus the following month.

Through his involvement over the years in events such as the air tours and air circuses, Cornish had developed a wide circle of solid friendships and professional associations within the aviation community. In a 1992 interview he clearly recalled the days when the members of the Indiana Aviation Association were encouraged to be goodwill ambassadors—and to have fun at the same time. "We'd fly our airplanes around to these various communities, find a farmer's wheat field or something to land on. They'd even close schools to show the kids our airplanes. I had a Curtiss Robin airplane the first time we went out, and I had Capehart Corporation sponsor this airplane.[59] A sign which hung on the flat side of the fuselage read 'Musical Marvel of the Age.' The person riding with me was Mary Bostwick, reporter for the *Indianapolis Star*. She was a great gal. These youngsters would get up there on my airplane. Who'd step out? Mary Bostwick. They'd kid her. Musical Marvel of the Age? She got the biggest kick out of that."[60] Such shared experiences created lasting bonds.[61]

But there were also travels of a more serious nature, some of which involved visiting with and testifying before legislators. In late 1933 or early 1934, Cornish went with Charles E. Cox and Harold Brooks, respectively the superintendent and the manager of the Indianapolis Municipal Airport, to Washington, D.C., where on behalf of the IATA they took part in discussions of the problems pertaining to aviation. The association expected its representatives to help legislators separate fact from fiction and gain a deeper understanding of the issues. The IATA and similar groups were troubled by the fact that Congress was crafting legislation that affected aviation even though the subject was alien to most legislators. One proposal would have had the federal government share the cost of a plane with the private operator (which would in turn give the government the right to purchase the title in case of an "emergency") and pay one-fourth of the cost of a student's training course, "thus allowing well-adapted and equipped training schools to continue to operate."[62] Cost sharing would encourage flying, some in high places believed, and would create a large and capable air reserve in place of the two or three thousand highly trained reserve members then available. A case of pie in the sky? It was definitely an idea that those in the industry did not want to see "fly."

By January 1935, there were new concerns among those in the aviation community about matters that they felt could have a negative effect both on them personally and on the broader field of aviation. One rallying point was legislation that would create a state aeronautics department. At a meeting in Indianapolis at midmonth, a delegation of airmen from South Bend and Fort

Wayne expressed their opposition to the idea. They feared that should the legislation pass, planes, aviation equipment, and the industry generally would be "severely" taxed in order to support the new entity.[63] Cornish called the proposed department "an unnecessary expense and an instrument which might become a victim of politicians."[64] As someone in a position to represent the view of many Hoosiers active in the field of aviation, he was assured by Fort Wayne's state senator, Chester K. Watson, that he would be granted a senate hearing if the bill went to committee.[65]

Cornish was armed with a new plan to submit, an alternative to the establishment of a state-level department. Drawn up in the form of a bill, what was referred to as the Cornish-Helmke aviation commission proposal called for the formation of an aeronautics commission for the state of Indiana, to be composed of eight members—two army officers and six civilians, serving gratis—who would suggest to the governor and the legislature ideas about the best ways to develop aviation within the state and coordinate those efforts with the work of the Federal Aviation Commission. It would also supervise the enforcement of aeronautical laws.[66] On one of Cornish's many trips to Indianapolis during the 1935 legislative session, which met from January 10 to March 1, he stopped to discuss legislative events with Frank Ball of Muncie, then IATA president, who expressed his opposition to a new state department and voiced approval of the alternative commission plan.[67] Neither proposal passed through the Indiana General Assembly that year.

The following year, 1936, was a quiet one for Indiana legislatively as the General Assembly met only briefly in special session from March 5 to March 18. But the IATA, wanting to be ready in the future with a well-considered approach, took action and appointed a committee of five "to investigate various forms of state aviation commissions and formulate legislation to effect one in this state." Cornish, Charles Bowers of South Bend, and Weir Cook, Charles Cox, and Walker Winslow of Indianapolis served on the committee appointed at the IATA's February meeting.[68] The process toward creating such a body for Indiana would be a slow one, but at least it was underway. Nearly a decade would pass before a commission as envisioned in the mid-1930s would become a reality.

Most of Cornish's flying during these years was civilian in nature, but he also developed personal relationships through the annual two-week sessions he attended at regional military bases. Although the U.S. government did not yet have justification for a boots-on-the-ground commitment in 1938, hostilities were growing worldwide, and it was necessary for the nation to maintain readiness through military reserve training. In December of that year, Cornish attended a twelve-day reserve camp at Schoen Field, Fort Benjamin Harrison, in Lawrence, Indiana. During this session, he and copilot Mike Murphy concentrated on instrument flying, radio instruction, flying radio range, orientation problems, cross-country navigation, and formation flying.[69] Such training was one more way for flyers to build friendships and allegiance to one another and to the country. It also helped prepare them for the very real possibility that they would be needed for a war that would be fought largely from the air.

Flight was becoming more routine by the 1930s. People were traveling by commercial aircraft in ever-greater numbers. Flying was safer, too, despite the occasional tragic accident, and conditions were far more pleasant given the improvements in engineering, technology, and aircraft design. But the growth in aviation was putting a strain on the nation's airports, and those who worked in the field were finding their jobs increasingly demanding. As the manager of a facility considered critical to the Fort Wayne economy, Cornish would be faced with a host of new challenges as he contemplated how best to guide the city's municipal airport into the future.

6

How to Grow an Airport

"Model airports aren't accidents. They don't, like Topsy, just grow up. They are the product of foresight, careful planning and plenty of hard work," said Cornish, speaking at the International Aircraft Show in Chicago in the winter of 1938. An airport manager "must be a politician, diplomat, salesman, publicity expert, showman, radio operator, bookkeeper, policeman, engineer and, incidentally, a trained pilot." Because Fort Wayne's airport was regarded as one of the most modern and best-equipped fields for a city its size, the Illinois Air Pilots Association had invited Cornish to share what had been done at Paul Baer Municipal Airport and what plans were being made for its future development. In his speech, he stressed to the assembled airport managers and commercial operators that before all else, "the public must be sold on the need of aviation facilities, the safety of flying and the commercial advantages of air transportation."[1]

Cornish was the second manager of Fort Wayne's municipal airport. He succeeded Captain Robert Bartel, who was appointed in 1929. Six years later, when the Board of Aviation Commissioners decided to change management, Bartel became the focus of dissension. He had built a strong following, and his supporters disagreed with the board's decision to hire Cornish. FERA (Federal Emergency Relief Administration) officials lauded Bartel's energy, efficiency, and general ability, while commercial operators at the municipal field charged that the board had "used the ax simply because he lacked what they described as tact [in interacting with the public]."[2] Though the board members admitted that Bartel had improved the field, erected hangars, and constructed runways, and while they praised his honesty, integrity, and ambition in building up the airport to its high rating, they went ahead with their decision. At that point the Fort Wayne Federation of Labor requested that their investigative committee look into the history of the firing, and the board agreed to work with them. If Bartel was found to be capable, the federation would insist that the decision be reversed. The investigation must have revealed the fact that Bartel had been reprimanded by the board several times for his failure to be courteous with the public. In the end, the federation determined that the board was within its right to replace him. On August 1, 1934, the reins were handed to Cornish, along with a monthly salary of $150.[3] He would hold the managerial position until 1941, when he was called to active duty in World War II.

The *News-Sentinel* praised the commissioners for hiring Cornish: "Possessed of an agreeable disposition and cheerful temperament, thoroughly familiar with aviation in both its technical and commercial aspects, Capt. Clarence F. Cornish may well be expected to give efficient and all-around satisfactory service in the discharge of the important new duties to which he has been assigned."[4] However, another publication commended Bartel and made this backhanded comment: "If what they [the Board of Aviation Commissioners] want is a toady to certain self-proclaimed 'big shots' they can find a more satisfactory candidate."[5]

Plans were underway for possible improvements at the airport when Cornish took over. Earlier in the year, the Airport Division of the state Civil Works Administration had developed a plan for new and improved airports and landing fields that would meet the requirements of the Aeronautics Branch of the federal Civil Works Administration (CWA). Fort Wayne was among the cities being considered for the allocation of poor relief funds to create jobs for some of the people left unemployed during those years of the Great Depression.[6] But even though there was little hope at the time for infrastructure improvements, and despite the naysayers, Cornish got off to a good start when in August he submitted a budget of $20,371 for 1935, which was lower than the previous year's budget by $1,500, coupled with a promise to economize in all departments.[7] Additionally, the cost of insurance coverage would drop because he was organizing an airport fire brigade composed of airport officials and attendants, pilots, and representatives of the flying services located at the facility. All were to be instructed in proper and effective firefighting methods by the Fort Wayne Fire Department, and any firefighting equipment that had previously been misplaced or forgotten would be restored to working condition. He also would immediately remedy a hazardous situation by directing workers to remove a motorcycle racetrack at the west end of the airport. And at no cost to either the city or the airport, Cornish would offer a public ground school course for $5 per student, paid to the city, and volunteer to be the check pilot for students before their first solo flight, or at any other time.[8]

The new manager would be expected to increase the airport's revenue by raising overnight storage rates for aircraft and decrease its expenses by examining the cost and efficient use of coal for heating. However, he would also have to deal with Hangar One, which was leaking in several places. If money could not be found for the needed repairs, the furnace and other vital equipment risked being lost.

For Fort Wayne's aviation commissioners, making the city's airport a place where the citizens would feel at home was a top priority. At a Christmas party in 1934, Cornish told the pilots and staff that he had been asked to have an ice skating rink built in the parking lot to provide wintertime recreation for the public, complete with lighting and benches. Beginning late in 1934, a dike less than one foot high was constructed, enclosing a 100-by-150-foot area that was filled with water by a city fire engine.[9] Bonfires would be built to warm fingers and toes. At least once, an ice hockey game was held there, with skating exhibitions between halves. Members of the winning team were promised rides over the city by the Inter-City Flying Service.[10] Inside the operations building, "a small but classy little restaurant with a two-by-four dance floor" was opened.[11] The *Journal-Gazette* had this to say: "To attract visitors to the field is worthy of favorable consideration since Indianapolis has a municipal airport that provides a pleasant place for the people to go for a sip, a bite of food, and perhaps an airplane ride and on summer evenings the Chicago field is a mecca for thousands."[12]

The public enjoyed driving out to the country for a pleasant afternoon or evening at Paul Baer Field, but there was one legal problem that purportedly affected the frequency of casual visits. Even though the idea of selling beer and even constructing a dance hall was presented by the Airport Board and Deputy City Attorney Howard A. Benninghof in July 1933, and despite the absolute end of Prohibition on December 5 of that year, Fort Wayne had passed a law banning the sale of beer outside the city limits.[13] The airport was municipally owned, but it was not located within the boundaries of the city, so "the amber beverage" could not be sold there for the public's enjoyment. The airport now had a restaurant, but the diners could not sip a beer with their meal. When business at the airport dropped off, especially on warm summer nights, City

Attorney Walter E. Helmke came up with a possible solution. If Fort Wayne could annex the airport, he said, beer could be sold there without a change to the existing law. For the sum of $5 a year, the New York Central Railroad agreed to lease to the Board of Aviation Commissioners a strip of land one foot wide and nearly two miles long that extended along the right-of-way between the city limits and the airport property. The last line of the legal document read: "We hope you get your beer."[14]

On another light note that year, the pilots at Baer adopted three flying squirrels.[15] The critters were captured in the woods north of the airport, where Works Progress Administration (WPA) workers were clearing trees considered to be a flying hazard (for planes, not squirrels!), and were kept in cages by pilot Whitney Gregg. The local aviators were surprised to learn that flying squirrels glide and suggested that pilots' licenses be issued to the unusual mascots.

Attention had to be given to the buildings and their upkeep, so in 1935 the commissioners chose to use FERA funds to build an office addition and dig a basement under the operations office where a furnace could be installed that would heat the entire building.[16] Also, bituminous mat paving was laid around the administration building and along part of the south side of the field on Ludwig Road.[17] Cornish was pleased and anticipated an increase in traffic when he and Transcontinental and Western Air (TWA) agreed to collaboratively install a traffic control system at the airport with a two-way radio communication set to be used for standby emergency purposes. "TWA would install a separate receiving set to maintain a constant watch on its planes while the airport sending set could be used to guide troubled planes in."[18]

While attending to the wishes of the airlines, local officials wanted to accommodate visitors and make it easier for them to park their cars at the airport, so they made plans to have pavement added that summer. They also wanted to be ready for increased usage of the airport's facilities when commercial passenger flights resumed—but as it turned out, that would take some time. On February 9, 1934, President Franklin D. Roosevelt abruptly canceled "all domestic airmail contracts,"[19] a revenue source that had been critical to commercial air operations. Daniel Rust calls this "the greatest upheaval the airline industry would know until the Airline Deregulation Act of 1978,"[20] As a result, no airlines were flying out of Fort Wayne, although limited airmail service did resume with the issuance of new contracts in May. Cornish, accompanied by senatorial candidate R. Earl Peters and Paul Guild, chairman of the Fort Wayne Chamber of Commerce Aviation Committee, flew the *Yankee Clipper* to Washington, D.C., that month to confer face-to-face with Postmaster General James A. Farley, but neither airmail nor passenger service returned to the city until 1937.[21]

Nevertheless, there was no shortage of flying activity at Paul Baer Municipal Airport, and regional events accounted for some of that. During the last week of July 1935, traffic increased significantly as people came to see airplanes from all over the country that were stopping at the airport on their way to or from the All-America Air Show being held in Detroit.[22] And in September, eight pilots who were competing for the Toledo Blade Silver Trophy landed their cruise planes at Baer and were served lunch before going on to Jackson, Michigan, the third lap of the course, which had begun in Springfield, Ohio.[23]

What community leaders wanted most of all for their city's stature and viability was commercial airline traffic. J. Ross McCulloch (whose first flight had been with local hero Art "Bird Boy" Smith), Paul Guild, Mayor Harry Baals, and Cornish worked valiantly toward this goal in the mid-1930s. They met in Chicago with Edward G. Bern, the president of Columbia Air Express Inc., hoping that the airline's tri-motor Stinson plane could add a stop each way on its daily route between Columbus, Ohio, and Chicago. Columbia needed airmail contracts to make a stop in Fort Wayne viable, and as there seemed to be no chance for that, the airline abandoned the idea. Unable to get past this sticking point on their own, officials turned to Senator Sherman Minton for help. Despite the fact that his office believed that Fort Wayne had excellent railway mail facilities, he promised that the legislature would include money for airmail service in the next round of deficiency appropriations. But that did not happen, either.[24] It was clear by April 1936 that Congress would not appropriate sufficient money for the Division of Air Mails to extend service to Fort Wayne. The division had operated under a deficiency appropriation the previous year, and that amount had to be added to the next fiscal year's request, meaning that it could not take on additional expenses.[25]

Keeping the citizens and aesthetics in mind, airport officials implemented beautification plans for the flight center while the major functional improvements were underway. In front of one of the administration buildings, Cornish added landscaping that included a waterfall of circulating water streaming down into a lily pond surrounded by shrubs and rocks.[26] The attraction "engendered a spirit of hospitality at the airport," reported the *News-Sentinel*'s Clifford B. Ward in his regular Abracadabra column.[27] It also would have been an attractive focal point had Fort Wayne been selected to host America's elimination trials for the 1935 Gordon Bennett International Balloon Race. Cornish, "who initiated the

Crowds gather to admire the airport's new landscaping. (Author's collection)

drive to bring the races" to Fort Wayne, worked hard to achieve that objective. He is shown in a newspaper photo shaking hands with Akron balloonist Ward T. Van Orman, who stated that he was "particularly impressed with the parking and traffic facilities, the railroad spur and surrounding topography."[28] Hopes remained high as late as March 18, but it was not to be.

Since Fort Wayne had not been successful in getting airlines to add scheduled passenger routes there, officials took a different approach. The airport would offer informational aid for pilots and improve the city's airfield so that it could be used for emergency stopovers. That included the addition of a modern communications system: a two-way radio. United Airlines had developed the first practical two-way radio, enabling pilots in the air to speak with people on the ground. For $750, Paul Baer Field added a two-way shortwave radio system made by RCA, similar to the equipment being used in Detroit, New Orleans, Chicago, Cleveland, and Battle Creek, Michigan.[29] Cornish passed a radio-telephone operators test in Chicago and was ready at the radio desk when the station, using the call letters WAJB, began operations on October 5, 1935. During the day, he shared the duty of maintaining radio communications with Earl Miller, secretary of the Board of Aviation Commissioners; two Indiana Technical College students worked the evening shift from 5:00 to 9:00. Thus WAJB could provide two-way protection and guidance for the eight TWA planes that flew the daily route from Pittsburgh to Chicago over Fort Wayne.

CORNISH AND MILLER OPERATING NEW RADIO STATION AT MUNICIPAL AIRPORT. Earl Miller (left) secretary of the Board of Aviation Commissioners, and Capt. Clarence F. Cornish, manager of the Paul Baer Municipal Airport, operate the recently-installed short-wave radio sending and receiving station WAJB in the office of the airport. Both are third-class licensed operators.

Cornish and Earl Miller operated WAJB, a two-way radio station that helped establish Baer as an emergency landing field. (Unidentified newspaper photo)

In poor flying conditions, Cornish and the others could direct any plane within a radius of 50 miles to the field; when flying conditions were good, their reach extended as far as 170 miles. In the first month after the radio equipment was installed, the airport's operators relayed current weather reports and other vital information to 130 pilots.

With the two-way radio in operation, TWA and United classified Baer as an emergency landing field. While the airlines were grateful for the added capability, they felt that more equipment would be even better. When TWA's John Collings made a plea for a radio range beacon, Cornish concurred. A beacon, he said, "is one of the newest developments in blind flying and is used by major airports throughout the country to direct air traffic in fogs, storms and rain . . . [and] would prove an inducement for ships to land here during inclement weather." When the airlines also let it be known that they would need greater landing capacity, Fort Wayne's aviation commissioners and Cornish responded quickly.[30]

The timing was perfect. President Roosevelt and the federal government were in a position to extend a giant helping hand—and a lot of money. In designating the week of October 14–21, 1935, as Air Navigation Week, Roosevelt proclaimed: "I particularly urge state and municipal officials, civic and commercial organizations, school authorities, and the press to call attention to the aerial-transportation achievements already accomplished, to set forth the advantages which may accrue as the result of continued development in this field, and to encourage the study of ways and means by which aerial transportation can contribute more effectively to the social and economic program of the human race."[31] On October 20, Fort Wayne's airport held an open house. The public inspected the new equipment, listened to plane-to-land communications through a public address system, and watched Harry Kaffenberg do a delayed parachute jump from five thousand feet.

If Fort Wayne was to meet the expectations of the airline industry, local and national factions would need to be on board. There was good news for the city's aviation interests when the federal government announced plans to double its $46 million allocation for airport projects around the country. The state of Indiana had previously received $1 million in WPA funds and therefore expected to be allotted $2 million the next time around. Cornish predicted that as one of Indiana's leading airports, Paul Baer Field would receive a large share of that money.

The Board of Aviation Commissioners wasted no time. In November it approved $278,000 for WPA-funded improvements. A plan was drawn up, and a map was created to show the extended runways, the new beacon, and the fifty-nine boundary lights that would replace the facility's outmoded coal oil lanterns.[32] W. B. Morgan, the WPA's project supervisor for Indiana airports, made a special trip to pick up the plans and confer with Cornish.[33] The project would have to be approved by a new division of the Bureau of Air Commerce, for which Morgan was the Indiana and Michigan representative. Cornish was elated when he learned that President Roosevelt had approved the WPA-funded project. "The project will be the biggest forward step in local aviation since the airport was started 10 years ago," he said. "These improvements have been a long-felt need at the airport. With the present out-moded lighting system, night operation has been greatly handicapped. Our asphalt runways also have been inadequate. This project, eliminating these two major flaws, will make the municipal airport a model field, and add much to the prestige of Fort Wayne as one of the leading aviation centers of the Midwest."[34] National airlines would finally have a reason to add Fort Wayne to their routes.

It was a major morale boost for everyone in Fort Wayne who cared about aviation there. Cornish was especially buoyed by the good news, and in the depth of the winter of 1935–36, he projected that the outlook for aviation in Fort Wayne during the coming year was the brightest in history. His optimistic forecast sprang from the anticipated funds, but it was also based on the fact that 1935 had been a very good year. The number of planes hangared at the airport had grown, according to his annual report, as had the number of students at local flying schools. Business at the repair shops was better than it had been since 1929. (The field had a government-approved repair station known as Consolidated Aircraft Repair Inc. and maintained planes from Fort Wayne and many far-distant points.)[35]

When the WPA allocated another $65,000 in the spring of 1936 to finance two new runways and extensions of two others, the *News-Sentinel* proclaimed that Fort Wayne would soon have the state's best airport. These were additional funds; the money for the boundary lights and the $9,000 radio range beacon had already been made available, and that work was already underway.[36] The beacon was functional by the end of May. It was located seven hundred feet north of the field and operated from inside the administration building. Cornish considered it a vital system and a boost to the facility's safety. "Through radio code, sent out in directed waves, the airplane pilot is enabled to keep on his course, despite poor visibility."[37]

By April, seventy-five men had begun digging ditches in preparation for the installation of the new lights. On June 23, Mayor Baals "threw the switch," brilliantly illuminating the airport as the city and WPA officials dedicated the most up-to-date boundary lighting system in the state. Cornish had urged the citizens to come out and inspect the modern facilities of their municipal field, and more than four thousand people were there. The officials were given flights and got a bird's-eye view of the airport's "new evening gown."[38] Thirty-five amber lights marked the boundaries of the field; seventeen red lights warned pilots of obstructions such as trees, poles, and wires; twenty-two green lights marked the

Electric lights replaced kerosene lanterns in 1936 to mark the boundaries of Paul Baer Municipal Airport. (Author's collection)

The asphalt used to pave Baer's runways was manufactured on site. (Author's collection)

approaches to the four runways. The runways were numbered and lit accordingly—the north-south approach having one light at the end, the northeast-southwest having two, the east-west having three, and the southwest-northeast having four.³⁹

Actual allotments would be slow to arrive, however. It was difficult to plan and to keep projects flowing, as funds arrived in chunks rather than in one lump sum and required ongoing negotiations.⁴⁰ Nevertheless, in June and July, workers hurried to complete the runways. Working in two six-hour shifts of seventy-five men each, they began construction on a new runway that would measure 2,200 feet long by 100 feet wide, running southeast to northwest. The job was expected to take forty-five days and use six thousand tons of asphalt. The plan called for the paving to be completed before cold weather set in, and certainly no later than the end of 1936. By November 25, the field had four runways totaling 10,600 feet in length.⁴¹

Yet despite the unflagging efforts of the airport staff, aviation commissioners, and public leaders, there were still no commercial airlines making regular use of the airport. Cornish was undaunted; ever the optimist, he directed his administrative crew to assist the public with any needed help or information about air travel. They provided fares, schedules, and even reservations for trips, though none of those trips originated in their home city. Airmail contracts, which provided the underpinning for the nation's commercial air traffic, remained out of

Fort Wayne's reach at the end of 1936. And until there was airmail service, there would continue to be no passenger flights.

In February 1937, Cornish and McCulloch took the train to Chicago, then flew as guests of American Airlines to the nation's capital to appear before the Interstate Commerce Commission and express their support for American's petition to establish an airmail, passenger, and express route between Indianapolis, Fort Wayne, and Detroit. The two men had with them a survey of business and industrial firms, conducted by the Fort Wayne Chamber of Commerce, to demonstrate the air traffic needs of their city. C. R. Smith, the president of American Airlines, emphasized his belief in the need for local air service between the three cities. Eddie Rickenbacker of Eastern Air Lines was there to challenge Smith. He contended that American had no intention of competing with any other airline in starting the new run and emphasized that the proposed routes were to be local only and would not infringe on others. The Interstate Commerce Commission sided with Rickenbacker and denied American's petition.[42] It was yet another disappointment for Fort Wayne's aviation and business interests.

The tide began to turn in May, when the Post Office Department awarded airmail contracts to TWA. Back in May 1931, the airline had begun making scheduled stops in Fort Wayne as part of its route from Chicago to Columbus, Ohio, but that service had been discontinued in April 1933 because landing conditions were considered too dangerous. However, things at Paul Baer Municipal Airport had changed greatly since then. A July announcement confirmed that TWA would have two airmail contracts, San Francisco to Winslow, Arizona, and Chicago to Dayton, and the latter included daily stops that would return air passenger, mail, and express service to Fort Wayne. The city would soon be part of the huge TWA transcontinental system.[43] There might finally be scheduled passenger flights at Baer.

In early August, John Collings, superintendent of TWA's Eastern Region, arrived in town to discuss the opening of air service. Cornish and Lois joined him for dinner at the Hotel Keenan, where Collings complimented Cornish and the members of the Board of Aviation Commissioners on the improvements made at the airport during the preceding years. He highly praised the foresight that officials had shown in making the decision to install radio range beacons and boundary and beacon lights even though no airline was operating in the city at the time. "The local airport is one of the finest in the middle west," Collings concluded, "and Fort Wayne can well be proud."[44] And so commercial aviation would again come to Fort Wayne.

The week following the dinner meeting, TWA took the Cornishes and nine community luminaries, including sisters Loretta and Helene Foellinger, on a half-hour "courtesy hop" over the city in one of the airline's fourteen-passenger Douglas Skyliners. "In the cabin of the luxury liner, guest passengers found everything for their comfort in the air. Deep, soft-cushioned seats enable them to recline, tilt back for a nap, or lounge. Passengers inspected the cabin at the front of the ship where the captain and first officer kept in continual contact with the airport by means of two-way radio equipment."[45] On August 25, some newspaper reporters also spent some time on a Skyliner, joining the crew on a survey flight "to do a careful study of weather conditions prevailing in the area covered by the route, the checking of locations of emergency landing fields, the taking of bearings of radio direction finders, and the testing of two-way radio facilities." TWA hostesses studied the territory so that they would be able to talk about local points of interest and scenic landmarks.[46]

The mood in the city was festive in anticipation of the big event. The Wolf & Dessauer Department Store, a major downtown destination, featured an impressive window display dedicated to TWA and "The World's Mightiest Fleet of Super-Skyliners." On the morning of September 1, 1937, passenger and mail service finally resumed in Fort Wayne after a four-and-a half-year hiatus. Cornish was on hand to welcome the twenty-one-passenger Douglas DC-3 *Kitty Hawk* when it arrived at Baer from Chicago. Along with a crowd estimated at

A window display at the Wolf & Dessauer Department Store welcomed TWA to Fort Wayne in 1937. (Author's collection)

more than a thousand, he celebrated the culmination of his hometown's diligent efforts to bring back commercial air service. The residents of Fort Wayne now had convenient nationwide air connections. After flying to Chicago, passengers could transfer to the larger, more elegant Skysleeper (a modified DC-2 with fourteen sleeping berths), then continue on to Los Angeles for $204.74, or to New York for $72.74.

However, trouble was stirring beneath the jubilation. Before TWA's first official flight into Fort Wayne, the company had made it clear that additional improvements would be needed, specifically space for both radio equipment and office personnel. The Board of Aviation Commissioners had requested $5,000 in July from the city's general fund for that purpose, but the City Council initially balked and then refused. On August 18, the council made a radical suggestion—the city should "sell the airport" and eliminate the annual budget line for its upkeep. With TWA preparing to include Fort Wayne on its route, officials on the council speculated, the cost to maintain the grounds and special equipment would soon increase. Therefore, why not offer the airport for sale to some private aviation corporation that might want to buy it? Without an airport to support, they concluded, an annual tax levy would be unnecessary.[47] Just as all those years' worth of effort were about to bear fruit, Cornish found himself on pins and needles.

Mayor Baals was not pleased. "This airport," he retorted, "represents an investment of nearly a million dollars considering what the city has expended there, plus funds the Federal Government has paid for its development through allocation of FERA, CWA and WPA accounts. Now that it has begun to show some chance of returning something on the investment, in addition to the pleasure it brings our citizenry, it seems to me any suggestion for its sale is devoid of any foundation in good judgment. The people want this field and they like to visit it—to feel that it is theirs and for their use, just as our parks are. Until I am convinced otherwise, I shall not consider selling the field."[48]

In August the City Council, in an executive session with a public hearing, proposed a two mill reduction in the following year's levy for airport purposes. The announcement caught aviation officials by surprise. The cut, which would save the average taxpayer all of eight cents, represented $2,000 that could not be applied to needed improvements to the airport. It meant that the administration building could not be expanded to meet the demands for more space created by the addition of TWA passenger and express service. Cornish had pointed out to the council that he and his secretary, Earl Miller, shared one tiny office with the TWA radio operator and ticket agent. Without the council's support, another improvement would have to be put on hold—grading seventy-eight acres of airport property north of the existing field so that the north-south runway could be lengthened from 3,500 to 5,300 feet. While a 3,500-foot runway had been sufficient in 1935, Cornish said, it was not sufficient in 1938. If the council had its way, the planned work to enlarge office space and prepare the airport for future commercial air traffic could not go forward. Gains that had been realized in recent years might be lost, he stressed, and the future of aviation in Fort Wayne jeopardized.[49]

The Aviation Commission countered with a proposal of its own, asking for a one cent levy that would raise $14,850, three-quarters of the airport's annual expenses. The City Council capitulated, and an impasse was averted. The council and the State Tax Board approved a payment of $5,000 to cover the sponsor's share of the $170,000 needed to expand the administration building and extend the north-south runway. A funding request was filed with the WPA. Ongoing improvements in aviation technology and expansion in the industry led to never-ending demands for upgrades on Cornish and his small staff, but with the financial assistance of one or another federal government agency, Fort Wayne and other communities were able to meet most of their capital project needs in the latter half of the 1930s.

In January 1938 President Roosevelt, in a message to Congress on national defense, cited the ominous fact that one-quarter of the world was at war, a hint of what was looming on the horizon for Fort Wayne and the nation. He asked Congress for an initial allotment of $800 million for defense spending, primarily to expand the navy.[50] International frictions and confrontations grew daily, as captured in Fort Wayne's newspaper headlines. As news came in February that the Spanish had sunk a British freighter and that Hitler had transformed Germany into a totalitarian state, the country's leaders grew increasingly concerned. "Demands for Gigantic Air Force Voice: Bloc Is Forming in Congress to Insist on More Planes for U. S. Defense," the *News-Sentinel* proclaimed on the 26th. "Airplanes can operate to protect our coast line up to a distance of 500 miles from shore. And while one battleship costs at the very least $50,000,000, a fleet of at least 500 bombers can be built for that amount."[51] Two weeks later, FDR's request for naval expansion reached $1 billion as "Hitler Makes Triumphant Entry into Austria."[52]

Fort Wayne and aviation interests would soon feel the impact of the world calamity in a very real way, but in 1938 it was still civilian rather than military aviation that had the city's attention. The year began well for Cornish and the municipal airport when *National Aeronautics*, the magazine of the National Aeronautic Association, featured Paul Baer Municipal Airport in its January issue. According to captions accompanying photographs of the flying field, the airport would need only minor extensions of the runways to meet the proposed new Bureau of Air Commerce standards for a Class 1 airport. Class 1 landing fields were required to have at least three paved runways, each at least 3,000 feet long, with a combined length of 13,500 feet. Paul Baer Field had four paved runways, but they fell 1,200 feet short of the total required length. The article encouraged Fort Wayne to make the necessary changes, even though the federal government might need to contribute generously to the cost. "Needless to say, there are not many airports which meet these requirements as to runways and effective lengths at the present time. However, if this amount of room is necessary, it ought to be provided, even though the Federal Government may have to participate very generously in the cost of further extension and development." It added, "Captain Clarence Cornish can be justly proud of the splendid service rendered by this field. Among the interesting features of the Fort Wayne airport are the very modern and spotless restrooms and the sleeping quarters available for transient pilots who may not wish to go downtown."[53]

Two months later, the WPA approved $33,450 for the desired grading work (in advance of paving) and expansion of the administration building to relieve cramped office conditions, provided that the air traffic control tower would also be enlarged to a room about fifteen feet square, and that the sponsor of the project, the Fort Wayne Board of Aviation Commissioners, would contribute $8,620 as its share of the cost.[54] Inevitably, bureaucracy again had a role to play. Before the project could be implemented, it needed approval from both President Roosevelt and Acting Comptroller General Richard Nash Elliott, and Indiana's WPA administration had to move the project to the top of its long list of federally approved projects and allocate funds from within its limited budget.

The value and impact of such work relief funds were inestimable. By providing paid jobs for millions of Americans who had been left unemployed by the Great Depression and the 1937 recession, the WPA helped ease the dire economic situation in the heartland and across the nation. In 1934–35, "public works projects, designed to take men off the relief rolls and put them on pay rolls," had brought $5,273,720 to Allen County.[55] The airport had been one of the chief beneficiaries of WPA and FERA funds in Allen County in 1934. Federal monies had been used to create a rock garden, add a heating system and brick veneering to the administration building, clear eighty acres of ground, till the field in preparation for runways, repaint buildings, and construct sidewalks. These improvements would have been unaffordable without the federal relief programs, Cornish said.[56]

With the coming of spring came praiseworthy news that gave a boost to morale. Major A. B. McMullen, chief of the Airport Section of the Department of Air Commerce, paid a surprise visit to Paul Baer Field and termed it "far superior to airports in other cities of this size. . . . It is better maintained and shows better management than most airports of the country." Having last visited around ten years earlier, he added, "I have found a better airport with better runways, buildings and equipment than I expected to find." However, he went on to say, "judging from the increase in traffic, it won't be long before you'll need a new hangar and administration building. As your airport is today, Fort Wayne probably will receive the second highest rating it is possible for the Department of Commerce to give." He praised Cornish for his accomplishments in developing airport facilities in the previous few years and was pleased to learn of the plans to construct an addition to the administration building and extend the north-south runway that summer.[57]

Continued emphasis on the role of federal assistance, coupled with the need for continually upgrading the nation's air program, came as a result of the Civil Aeronautics Act of 1938, which was signed into law in June. It required the Civil Aeronautics Authority (CAA) to submit "a detailed nationwide survey of airports" to Congress. According to the FAA (Federal Aviation Administration),

> the report indicated that the number of municipal and commercial airports had increased from 823 at the end of 1927 to 1,833 at the end of 1938, and that Federal relief programs had been responsible for most airport development since 1933. The Authority recommended that the development and maintenance of an adequate system of airports (including seaplane bases) should be recognized as a matter of national concern and a proper object of Federal expenditure. Currently, the Authority believed that airports should receive $100 million of regular public-works or work-relief funds, as well as $25 million to increase the Federal share of joint Federal-local projects.[58]

With airmail and passenger service in place and construction projects moving along as needed thanks to federal funding, the autumn of 1938 was relatively calm for the airport's manager—a good time to relish the birth of his first and only child, his daughter Ruth Ann, born September 4, and to bask in the attention surrounding a four-day visit to the airport by the "Famed Transatlantic Flier" Clarence D. Chamberlin.[59] On October 20, Chamberlin, Cornish, and Paul Guild, vice-president of the Board of Aviation Commissioners, reunited in Fort Wayne, most likely reliving their times at Wright Field and in the Army Air Corps' pilot training program during World War I. Chamberlin—accompanied by his wife, Louise Ashby Chamberlin, who was said to be the "former air companion of Amelia Earhart"—had gained fame for completing the first transatlantic passenger flight only two weeks after Lindbergh's epic solo journey, flying nonstop from New York to Eisleben, Germany, in a little over forty-three hours. Chamberlin's Fort Wayne stop was part of a tour to raise funds for research into the efficacy and safety of air travel, especially stratospheric flights. Among his goals were to develop a plane that could fly at a speed of at least four hundred miles an hour and to be able to cross the Atlantic in four hours. While he was in town, he offered Fort Wayne residents the opportunity to fly with him in his twenty-seven-passenger Curtiss Condor plane.

On the evening of Chamberlin's arrival, Colonel Roscoe Turner was the guest speaker at the Fort Wayne Engineers' Club, for which Cornish served as toastmaster. Four hundred people gathered at the Catholic Community Center that Thursday night, with members of multiple community organizations in attendance, and with Chamberlin the event's guest of honor. Turner stressed the importance of airplanes in peaceful pursuits as well as in national defense. "If France and England had been prepared in aviation, they'd have never said, 'Yes' to Hitler."[60] Was it a coincidence that both renowned aviators appeared in Fort Wayne at the same time? Or did one's visit inspire the other's?

City and aviation officials were elated when, on December 15, 1938, TWA added Fort Wayne to its nighttime route between New York and Chicago. "Everyone interested in aviation and in the inauguration of night flying" was "extended an invitation to attend the celebration at the airport" when the westbound airliner through Dayton was to arrive on its way to Chicago.[61] This facilitated cross-country passenger, express, and mail service, which were "highly important to a center like Fort Wayne, strategically located as a major city in the center of the east and west transcontinental traffic."[62] Of note, passengers could fly from Fort Wayne to New York in only four hours and thirty-three minutes.

The new year began on a discordant note internationally as world peace continued to disintegrate. Yet while the president maintained a position of neutrality for the United States, he was cognizant of the possible consequences for the nation. A *News-Sentinel* headline four days into 1939 declared, "Roosevelt: Lashes at Dictators: Warns of Menace to Freedom; Must Be Ready to Fight in Military and Economic Sense; Preparedness Urged." Eight days later, FDR asked Congress to approve half a billion dollars for defense.[63]

For most in the United States, though, life went on as usual, and Cornish decided to attend a January conference in St. Louis, a timely decision that put him in touch with men and ideas beyond Indiana's borders. He emerged from the meeting as the eight-state regional leader of the National Association of Airport Executives, which co-sponsored the event with the National Aeronautic Association (NAA).[64] Back home, Cornish met with aviation board members, educators, and model builders (the Mad Modelers Club) to form an NAA chapter, later named after Art Smith. Providing aviation programs for boys in public schools was seen as vital for preparedness in air-related vocations, and in March, at the first organizational meeting, the group clarified its purpose—to supervise and maintain an aviation program of education and recreation in grammar and high schools. (When the group had become more fully established, its purpose statement was expanded: "to call attention to the growing importance of aviation in the national life.")[65] Wasting no time, Cornish organized an airport forum for May, with "many of the best-informed aircraft officials in the country" expected to attend. Four hundred invitations were mailed to airport managers in NAA's Region 3, for which he was director. The purpose of the forum was to "iron out difficulties besetting airports and to further the cause of efficient management." The topics and speakers were as follows:

"Airport Planning and Development," Colonel Floyd E. Evans

"Airport Lighting," Nish Dienhart, Manager of the Indianapolis Municipal Airport

"Federal Aid to Airports," Major C. V. Burnett, Manager of the Detroit City Airport and President of the American Association of Airport Executives

"Management and Regulations," B. E. Fulton, Manager of the Akron Airport

"Airport Zoning," Judge George H Leonard, Attorney for the Board of Aviation Commissioners

"Airport Surfacing," a joint presentation by William Aldous, CAA Airport Engineer, and W. R. Macatee of the Asphalt Institute Inc.

"Airport Traffic Control," Major J. H. Berry, Manager of the Cleveland Airport

"Airplane Hangars (types and methods of financing)," Cornish[66]

The forum, which was attended by seventy-five managers from Region 3, emphasized the need for federal funding to support the nation's airports. "There is no Federal aid of airports," Burnett declared, "but managers should demand a subsidization by the Government." He noted that there were approximately 2,200 municipal and privately owned airports in the country that could accommodate interstate commerce activities—the key word being *interstate*.[67] The regional airport executives who had gathered for the forum were therefore pleased with the CAA's recommendation that $125 million be allocated for lighting, runway construction, and navigational facilities such as control towers, with the needed land to be provided by the relevant municipalities.

Ten days after the resoundingly successful forum, there was a stellar development in the history of Paul Baer Municipal Airport: it was tapped to accommodate a U.S. Weather Bureau observation branch.[68] It was a real coup for Fort Wayne, as it would put the city on a par regionally with Chicago, Cleveland, Indianapolis, and Detroit. The branch would open on January 1, 1940, staffed by ten full-time employees, five CAA men, and five Weather Bureau observers, with an annual payroll of $20,000. The article concluded: "The bureau to be established here is the Class I type and would have few superior to it in the country."[69] Federal money was crucial and deemed totally appropriate; it would be used to purchase the instruments and equipment, while Fort Wayne would pay for an eight-hundred-square-foot addition to the south side of the terminal building. Cornish had plans ready by the end of the day on May 25, so that bids could be opened and the exact cost determined. Initially noncommittal regarding the idea, the City Council ultimately approved the expenditure.[70] The final cost was in the $3,500 to $5,000 range.

While these positive developments moved through the bureaucratic pipeline, and while daily life continued as normal for most in Fort Wayne and throughout the nation, ominous reports from overseas were arriving with ever-increasing frequency. On September 1, a *News-Sentinel* headline boldly announced, "Undeclared War Starts: Roosevelt Says U.S. Will Stay Out." There was no doubt about the reality of war for Londoners, however; the next day's paper reported that as a precautionary measure to protect the residents of that city should they be attacked by the Germans, "all venomous snakes, black widow spiders, bird-eating spiders, scorpions and other dangerous insects of the London Zoo" had been destroyed overnight.[71] But later that same month, Fort Wayne and other U.S. cities with NAA chapters participated in National Air Week, or National Air Progress Week, which was launched on September 18. The Art Smith Chapter of the NAA, the Aero Club, the Mad Modelers, the authorities at Paul Baer Airport, civic organizations, and municipal and federal officials all took part.

At the airport, the public got to visit the newly completed administration building annex, dedicated during that week, and the Anthony Wayne Stamp Club provided an exhibit of airmail stamps. The Mad Modelers held a "gasoline model airplane contest," and they also had full-sized planes on display, with signs identifying the type, weight, power, and purposes of each. Downtown stores featured special window displays. One of the high schools had a daily showing of "talking motion pictures" furnished by American Airlines and Pan American Airways. Two movies were featured: *The Rio Cruise*, which showed an actual flight from Miami, Florida, to Rio de Janeiro via Cuba, the east coast of South America, and Brazil, "all accompanied by music and a narrator"; and *The American Way*, a film so realistic that it made viewers feel as though they were actually flying in a plane.[72]

Believing that aviation centers should be interconnected, and wanting to make regional air travel easier and more convenient, Cornish called a November meeting in Indianapolis with state, federal, and local aeronautics officials to discuss a four-state air highway system that would link the Indiana cities of Fort Wayne, Frankfort, Indianapolis, Marion, Muncie, and South Bend with comparable cities in Illinois, Michigan, and Ohio through regularly scheduled runs—a "secondary 'feeder' system."[73] Cornish saw this as a model for the nation, "in harmony with current aeronautical plans to expand the facilities of municipal and private airports in America." It was noted in a follow-up article about the meeting that the group considered drafting a resolution to Congress urging that funds be appropriated for an extensive expansion of the nation's aviation program.[74] Plans such as these being deliberated for civil aviation in the late 1930s would soon be interrupted by the war, but in the mid-1940s they would become a significant portion of Cornish's and the state's endeavors.

But for the moment, the big aviation news revolved around the Weather Bureau branch, which became fully functional in January 1940. The crew began giving hourly reports by teletype on ceiling visibility, sky conditions, temperature, humidity, barometer reading, and wind direction. *News-Sentinel* reporter Charles Keefer was engrossed. "Like silent sentinels," he wrote, "the Weather Bureau

Airport Stations and the Civil Aeronautics Authority Communications Stations . . . are truly the 'unsung heroes' of the Nation's vast network of air lines and many a pilot and air passenger owes his life to the constant vigilance of these two Government agencies." Keefer's article, published in a Saturday edition in February 1940, explained the inner workings of these agencies and described in detail the work of the men at Paul Baer Municipal Airport who kept an eye on the weather and transmitted vital information nationwide to maintain safe airways.[75]

The new branch's balloon service was one of the Weather Bureau's most important tools for measuring atmospheric conditions. Small balloons, three feet in diameter, were inflated with a lighter-than-air gas such as hydrogen or helium and sent aloft. Then a theodolite, an instrument resembling a surveyor's transit, followed their rise and oscillation or horizontal movement.[76] As twenty-first-century hot-air balloonists know well, the wind at one altitude may be in one direction with one velocity, while at another altitude both the direction and the velocity may be different. This information was invaluable both to pilots who needed to keep on schedule and to meteorologists making daily weather forecasts. As winds aloft can quickly alter conditions at ground level, such knowledge was essential to correct forecasting.

After years of dedication and advocacy by the Aviation Commission, Cornish, and the airport staff, Fort Wayne had gained well-deserved prominence as an important state and national aviation terminal, as a principal stop on TWA's Chicago–New York route, and as a center of vital information for civilian, commercial, and military pilots. The city paid off its ten-year, $100,000 bond issue and received clear title to its aviation center, appraised then at $1.2 million. When Cornish was asked his thoughts about an appropriate way to celebrate, he suggested that the retired bonds "be tossed onto the bonfire at the Allen County Court House mortgage burning [January 1, 1940] or scattered to the four winds from an airplane over the port."[77]

At that point, Cornish understandably could have envisioned a smooth course ahead. But 1940 was barely underway when a controversial issue arose and ignited a heated legal debate. A 32.08-acre piece of property had gone into receivership, and in January the Citizens Trust Company asked the Zoning Plan Commission to allow the acreage to be platted for a housing complex, to be called the Airview Addition. The problem was that the tract was directly across Lima Road from the southern tip of Paul Baer Field's southwest-northeast runway and directly underneath the primary flight path. The Board of Aviation Commissioners opposed the idea, claiming that having buildings so close to the runway would constitute a hazard. They believed they could fall back on a 1929 law that allowed them to make determinations about the use of properties adjacent to airfields. "If we are going to permit houses to be erected on this tract and other property in the immediate vicinity, we might as well use the airport to fly toy models and get an entirely new field," Cornish said. He felt strongly that approaches to runways must be kept clear, and that if the city officials should fail to do so, the city would be remiss in protecting its own investment, then in excess of $1 million. TWA officials weighed in, too: "This will naturally result in the installation of telephone poles and lines, etc., adjacent to the airport which will be hazardous to aircraft taking off and landing at Fort Wayne. This will also result, no doubt, in some of the people occupying this space complaining of planes taking off over their homes." The City Plan Commission concurred and denied the request.[78]

The trust company appealed the case to the circuit court, and despite the safety concerns and objections voiced by multiple parties, in June 1940 Judge John F. Decker of Bluffton ruled in favor of the landholder. Decker held that the police power of the 1929 Indiana General Assembly's airport statute did not give municipal authorities the right to prevent the landowner from developing the acreage, and if the city wanted to control the land contiguous to the airport, its only recourse was to condemn the tract or to purchase it. He told the attorneys that despite an exhaustive study, he had found no established law related to the subject to use for guidance. He pointed out that at one time property owners had been presumed to control both their land itself and the space above it. However, "aviation has changed this situation and the welfare of mankind requires that the rights of the owner of the surface be limited to his use and enjoyment of only so much of the space above his land as may be used for airplanes as long as they do not interfere unduly with the enjoyment of owner of the surface."[79] Simply put, according to Judge Decker, in this case condemnation or purchase would be preferable to invoking police power. The airport budget for 1941, submitted in the late summer of 1940, included $6,000 to purchase the 32.08 acres of land that had been designated for the proposed Airview Addition. This budget item was approved, but the City Council subsequently declined to appropriate the necessary funds. On March 3, 1941, with the issue having come back before them, the City Plan Commission had no choice but to approve the platting of the Airview Addition.

7

A New Baer Field and a Struggling Old

WHILE THE AIRVIEW ADDITION was in litigation, Cornish, on behalf of a citizens' committee, submitted to Fort Wayne's civil engineers a map he had sketched showing what the municipal airport might look like if it could be expanded to nearly twice its current size. The engineers refined the plan, with their first version dated April 25, 1940, followed by a revision dated August 10, 1940. Since it was economically essential for the airport to be able to accommodate the growing demand for passenger, mail, and cargo service, the city planners and the Board of Aviation Commissioners approved an expansion that would add 330 acres south and west of the property, enabling the addition of the new runways and hangars that would be needed in the coming years.[1]

Not only were the numbers of takeoffs and landings increasing in 1940, but commercial aircraft were becoming larger and heavier, demanding longer and sturdier runways. According to an article in the *Journal-Gazette*, a survey conducted by the U.S. Civil Aeronautics Authority (CAA) and published in "the current issue of 'American Aviation'" ranked Paul Baer Municipal Airport twentieth on the list of the nation's busiest airports (tied with the Chicago Airport), with 2,370 separate landings and takeoffs recorded during a test week the previous June.[2] According to Cornish, however, that result was somewhat misleading: "The majority of the two thousand or more airports throughout the Nation are operated on a 24-hours-per-day basis," the article quoted him as saying, "while the Paul Baer field operated on a 10-hours-per-day schedule during the test week and is now operated on a 15-hours-per-day schedule. Since the switch [to 15 hours], during the second week of August, 3,900 landings and takeoffs have been recorded."

Given this reality, the condition of Paul Baer Field's runways was a concern for the airport's manager and the Board of Aviation Commissioners. Most recently improved by the WPA (Works Progress Administration) in 1936 and 1937, with a life expectancy of five to six years, the runways were deteriorating; but reconstructing them so that they could handle increasingly heavy usage would be costly. Enter the knights in shining armor. To the surprise of many, the U.S. Army Corps of Engineers came to Fort Wayne in mid-August and expressed an interest in using Paul Baer Field for military training in army air transport craft. It would be a small Army Air Corps unit, and the training planes could operate alongside the civil aircraft that used the facility.[3] For the city leaders, the timing could not have been better.

If Fort Wayne's municipal field were to be selected as the site for an Air Corps unit, the federal government would presumably pay for the field to be enlarged and developed for both civil and military operations. It was front-page news in the local papers. The *Journal-Gazette* laid out the details of a joint announcement made by Cornish, as airport manager, and Charles L. Biederwolf, the secretary-manager of the Fort Wayne Chamber of Commerce. Fort Wayne might "become an important middle-western base for the aerial operations of the United States army."[4] However, South Bend, Indianapolis, and Evansville were also competing for this prize, so time was of the essence in deciding whether the city could accommodate the Army Air Corps.[5]

A letter was mailed to the city's leading citizens, inviting them to attend an evening event at the airport on August 20. The purpose of the meeting was to launch an aviation development program for Fort Wayne, the highlight of which would be a broad civic effort to secure a new military air base. Addressing the assembled group of fifty business and civic leaders, principal speaker Cornish and Ross McCulloch, president of the Board of Aviation Commissioners, made a case for the pressing need to buy the additional 330 acres of land and increase the size of the airport so that Fort Wayne could benefit from the expansion of national air defenses and the rapid growth of the aviation industry.[6] The group agreed in principle.[7] John W. Crise, an accountant for the General Electric Company, was named to head the Citizens' Committee on Aviation, which took the lead in promoting Fort Wayne's new aviation program.[8]

Time was not wasted. On August 29, Cornish was on his way to Washington as the group's envoy, where he met with Indiana senators Frederick Van Nuys and Sherman Minton, Congressman George W. Gillie, and officials of the CAA and the War Department to promote and gain support for Fort Wayne's bid to be the location for one of the proposed new bases for the Army Air Corps. Back in Fort Wayne, he reported on September 3 that the city had "every right to expect its share of defense projects in due course of time."[9] Three days later, Cornish returned to Washington, along with Mayor Baals, Indiana Service Corporation president W. Marshall Dale, U.S. District Attorney James R. Fleming, and Major Earle W. Moss, a reserve officer in the Army Air Corps. Their goal was to convince both civil and military officials that Fort Wayne was a strategically located city that offered numerous advantages.[10]

If column space in the *News-Sentinel* and the *Journal-Gazette* is a valid indicator of the significance of a particular topic, then attracting the Army Air Corps in 1940–41 would be near the top of the city's list. In the end, Fort Wayne would pay $125,000 for 728 acres of flat farmland so that a new airfield could be built southwest of the city. But first there would be months of wrangling.[11]

On September 21, officers arrived to inspect the proposed air base site.[12] By September 28, it was learned that the federal government had promised $399,500 for improvements to the airfield, which according to Cornish had been selected along with "28 other airports as a base for future extensive tactical operations."[13] Despite some concerns, the Citizens' Committee on Aviation, whose numbers had risen to 150, adopted a resolution favoring the acquisition of land for multipurpose use (including by the military).[14]

On October 11, the newspaper headlines proclaimed success, reporting that there was now a proposal for a base to be built at the municipal airport for 175 officers and 1,700 soldiers from the Thirty-First Pursuit Group. The amount to be invested had increased to $1 million. The article itself took a more cautious approach. It stated that Cornish had received no *official* information, and questions had surfaced about how these developments would affect Paul Baer Field.[15] "Should the Army decide to use the airport for its air base," said the next day's *News-Sentinel*, "it is likely that all civilian and commercial flying from it would be suspended." Cornish, however, expressed the belief that TWA would continue to stop at Paul Baer Field after the base was established.[16]

Even though the War Department had officially released plans for the air base only to the press and to congressmen in Washington, and not yet directly to Fort Wayne's stakeholders, the citizens remained generally optimistic, their unresolved questions notwithstanding. The city itself was rejoicing:

It should be a cause for extending heartiest congratulations to Capt. Clarence Cornish, to the Citizens' Committee headed by J. W. Crise and C. Byron Hayes, and to all other individuals and institutions who worked and planned for this great event.

For all of us are to benefit—militarily, industrially, commercially, and culturally—by reason of the fact that a few public-spirited leaders took it upon themselves to labor long hours for this to come to pass. . . . Our respected community builders of 1940 are as worthy of the esteem and even veneration of us all as are those who transformed Fort Wayne from a territorial stockade into the makings of a progressive industrial city.

Realizing these truths, proud and vigilant and ever constructive, we come to realize that—to those who have the will to win—all things are possible.

This is a great day in Fort Wayne history.[17]

Still, the situation was not entirely rosy. Mayor Baals wrote a letter to the War Department on November 2 to notify them that the city would need to issue a municipal bond in order to purchase the acreage for the new base—a process that would take ninety days.[18] Details of this potentially massive project remained in a state of flux. The following day, the press announced that the army had increased its appropriation to $1,378,000 and nearly doubled the anticipated number of personnel to be assigned to the base, from the original 1,700 to 3,000, by adding the Thirty-Ninth, Fortieth, and Forty-First Air Unit squadrons.[19] At that point, however, the city officials had yet to receive formal confirmation from the War Department.

The military had a gripe of its own. It claimed that after several weeks, it still had not received a response to the draft of a lease that it had sent to Colonel F. H. Boucher, the officer in charge of Army Reserve affairs in the district, to be negotiated with the city for 60 acres of land adjoining the municipal airport on which barracks could be built for enlisted men. However, the signing of that lease had necessarily been delayed because that acreage was part of the 330 acres that the city did not yet own, for which an as yet unissued municipal bond would be required.[20]

On October 11 it had been announced that CAA training programs would no longer be permitted at Paul Baer Field and twenty-eight other airports, because "military authorities have based, or plan to base in the near future, extensive tactical operations on these sites." Future CAA training was to be "conducted at an airport at least six miles from the present site," although it was "not certain when this ruling will go into effect."[21] Needless to say, the possibility of losing the pilot training program caused consternation among those committed to helping men and women learn to fly. Grove Webster, the director of CAA's student pilot program, promised to fight to keep it. "The Army is driving CAA from Paul Baer Municipal Airport at Fort Wayne," he said, "and it is up to Fort Wayne to provide facilities to continue the program."[22] On November 3, CAA engineers were in Fort Wayne to explore possible sites for the construction of a new airport so that their pilot training program could continue. They also surveyed the land that would be needed to expand the municipal field for the military base.[23]

Weeks went by, and there was still no official word regarding the military base. City officials remained in the dark, and queries sent by Baals to the War Department went unanswered. By late November, more citizens were registering concerns about the disadvantages of the proposals on the table. They became more vocal, wanting to make sure that their views were heard alongside the supporters' views. Ten representatives of local flying services were among the signers of a petition that opposed the base. If the federal government were indeed to request, and receive, permission for use of the airport, it would be a detriment to civil aviation in Fort Wayne, and likely would cost the city its air transport connections. "Any problematical business gain accruing to Fort Wayne businessmen would be completely offset, if not more than offset, by the crippling of civilian aviation services," opined the *News-Sentinel*.[24]

Behind the scenes, the possibility was floated that an alternative site might be chosen, a shift with the potential to allay the concerns held by many. Mayor Baals, having conferred with Cornish, revealed that he was considering a new location for the air base. He had listened to residents and taxpayers and local flyers and felt that their concerns about the potential loss to civilian aviation were justified. When the *News-Sentinel* announced this turn of events on November 26, it added that Cornish had been informed during a long-distance phone call that relocating the project "seemed feasible" to the army engineers, though they wanted to reserve the right to use Paul Baer Field temporarily, for about three months. Should the new site be agreed upon by all parties, the city would drop proceedings to acquire the additional acreage.[25]

Shortly after midnight on December 10, Mayor Baals received a telegram from the Army Air Corps, which read: "Request decision of city to procure new airport site and other installations." Baals responded by telegram, "Fort Wayne accepts Government offer to construct air base here." It was official, formal, and final. Said the editors of the *News-Sentinel*, "We are happy that Fort Wayne has generally been blessed with a prevailing deliberateness of judgment which, in the case of entire communities as in the case of individual citizens, is a highly serviceable attribute involving a maximum of substantial advantage in the long run."[26]

City officials were elated, but as the reality of a proposed air base began to sink in, the citizens of Fort Wayne weighed the pluses and minuses, among the latter being the number of new people who would come along with the base. The initial plan called for three plane squadrons, totaling nearly 60 aircraft, which would require 1,500 enlisted personnel plus 150 Air Corps officers, as well as their wives and families.[27] At local informational hearings, citizens voiced their opposition to the new base and expressed some concern about the character of the men who would be stationed there. Cornish rose up at least once in the airmen's defense: "Only the highest type of man mentally and morally is permitted in the air corps," he told them. "Officers must have a minimum of two years of college education, and even enlisted men must be at least high school graduates."[28]

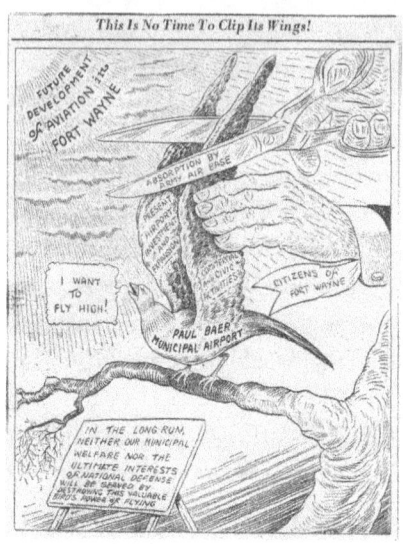

When this cartoon was published on November 28, 1940, some feared that the city might lose its bid for the air base. Two weeks later, Fort Wayne was officially awarded the base. ("This Is No Time to Clip Its Wings," *News-Sentinel*, November 28, 1940, 6. Used with permission from the [Fort Wayne, Ind.] *News-Sentinel*.)

Before the year's end, the military demanded that the lease for the land be settled and signed by February 1, 1941, or it would move the pursuit units out of Fort Wayne and relocate them elsewhere. During this highly charged time, Cornish again acted as a liaison between the city administration and the War Department. On January 4 he sent a telegram to Mayor Baals from Washington, reiterating the urgency of the matter from the military's perspective. Numerous conferences followed, mostly by telephone, during which the mayor "put it up to the business men."[29] Hasty arrangements were made to secure enough money to pay landowners and cover other costs; a total of $125,000 was needed to close the deal. Four Fort Wayne banks agreed to lend the money to the city

Baer Field was rededicated on September 21, 1991. Left to right: Roger Myers, president and curator of the Greater Fort Wayne Aviation Museum; Cornish; Robert Schott, the manager of Smith Field during World War II and founder of the GFWAM; Roxanne Butler, public relations manager for the Fort Wayne International Airport. (Photo courtesy of Roger Myers)

through a trusteeship, with repayment to be made through the issuance of a municipal bond. A group of twelve to fifteen prominent businessmen would serve as underwriters. Other Indiana cities were hopeful that a glitch of some kind would derail Fort Wayne's bid. Two or three cities reportedly were ready and willing to step in and find the needed funding should Fort Wayne's efforts fail for some reason.[30] In only two days, however, the city had the money and the War Department had its deal. On January 20, 1941, city and federal authorities signed the lease for the city's new airport southwest of Fort Wayne, to be used as an Army Air Corps base for a term of one to twenty-five years. Construction was well underway within a month, and the facility was ready for use by December of that year. It would continue to grow in the coming years.[31]

During the acquisition and construction phases for the new site, the military had the right to land and hangar its planes at what news articles were referring to that spring as Smith-Baer Field. When Colonel Eugene A. Lohman, the commander of the new field, first landed his formidable-looking B-18A twin-motored bomber at the old field on April 13, 1941, it caught the attention of the press and citizenry.[32] Commercial interests had been promised by the CAA that within a few months there would be solid concrete runways at the new site that could accommodate commercial aviation's increasing loads, but during that spring of 1941, they would have to share the crumbling old runways with planes such as Lohman's bomber. When the federal government poured the concrete for the runways at the new field, however, it did not add aprons and taxiways for civil aviation. Congress had appropriated $631,000 for the project on October 9, 1940, with the understanding that it would serve both civil and military purposes, but the allotted money had proved insufficient to cover all of the projected construction. Even though the military's plans, approved back on December 17, 1940, had designated a portion of the field for civil aeronautics, everyone involved, including Fort Wayne civilians, agreed to temporarily postpone the construction of taxiways and parking aprons for civil use, "with the clear and distinct understanding that these facilities would be provided for from future Federal Airport appropriations."³³

Colonel Eugene Lohman landed his B-18A bomber at Paul Baer Municipal Airport in April 1941. ("Col. Lohman's Bomber Arrives," *Journal-Gazette*, April 13, 1941. Courtesy of the *Journal Gazette*, Fort Wayne, Ind.)

As of August 1941, provisions for commercial flights at the new field were still not underway. It seemed that federal officials wanted exclusive use of Paul Baer Field for military purposes, "without the inconvenience of the use of said field for civil aeronautics."³⁴ This must have been a grave disappointment to Mayor Baals, Cornish, and the other civic leaders who had expected to transfer the heavier commercial air traffic to the new airport, with its properly designed runways and amenities. Had that happened, the city would have been able to limit the use of the municipal airport to light civil aircraft that would require only normal maintenance and repairs. The city was unprepared to pay for the upgrades that would be needed if the facility was to continue serving all types of aircraft. If TWA and the other commercial users had to face another winter without access to Baer Field's concrete runways, Fort Wayne leaders feared losing all commercial aviation business.

Mayor Baals carried on the appeal in the months that followed, asking the CAA to reallocate funds from Baer to Smith and to pay $150,000 for materials so that WPA workers could repair the asphalt runways.[35] These matters were not to be resolved during Cornish's tenure. On June 14, 1941, the newspapers announced that Cornish would be returning to active military duty on July 5.

In conjunction with the construction phase, a name change was contemplated for the new field, but this too was not a simple process. The War Department left the decision up to the city, since the military would only be leasing the property. In February the names of two important men in the city's history were suggested—General "Mad" Anthony Wayne, pioneer soldier and Revolutionary War hero, and Art Smith, Fort Wayne's pioneer pilot. On February 20, the Board of Aviation Commissioners announced that the new site would be called Anthony Wayne Field, and the old site would become Smith-Baer Field. Two weeks later, a new War Department ruling stipulated that the base must be named after a deceased Air Corps officer.[36] By mid-May, the Board of Aviation Commissioners had made its decision: the air base would be named Paul F. Baer Field. The military gave its nod of approval on June 9.[37]

A new name was now needed for the municipal field, and it was a matter that Cornish had strong feelings about. Since the late 1920s, he had advocated for a more appropriate way to honor Smith than with the stone pillar in Memorial Park.[38] One of his final acts as manager of Fort Wayne's municipal airport would be to strongly recommend to the Aviation Commission that they name the city's historic aviation center in honor of the beloved pioneer "Bird Boy."[39] In July 1941, the facility became Art Smith Municipal Field.[40]

Hangar Two at Art Smith Municipal Field was added to the National Register of Historic Places in 2002. (Author's collection)

Along with Cornish and millions of others, the citizens of Fort Wayne were rallying to the demands of a new war effort. At 9:00 a.m. on July 24, 1941, the members of the Aero Club took to the air and began flying in formation over Fort Wayne. The purpose of their flight was to remind the city's residents about the national shortage of aluminum, which was needed for manufacturing planes and other defense purposes. They were asked to place any aluminum pots and pans that they could spare on their front porches, along with old aluminum utensils and any other objects containing aluminum. "An old vacuum cleaner, a discarded serving tray or salt shaker, pots and pans. . . . Father's old automobile engine, brother's old bicycle. . . . All aluminum contributed will be used exclusively for national defense."[41] A caravan of fifty trucks, manned by drivers and Boy Scouts, then drove from house to house to pick them up. The Aero Club members were glad to have such a visible role in the effort. Like the public at large, they sensed the gravity of what lay ahead.[42]

By then Cornish was in Louisiana, where he would be expected to apply his civilian and military expertise to the United States' rapidly evolving participation on both the

> One of America's oldest surviving early aviation sites, Smith Field was added to the National Register of Historic Places in 2002. Its three buildings remain in their initial configuration and traffic has increased following a reconstruction project in 2009 and 2010, leaving both runways in "excellent condition."[43] "Hangar Two, built in the 1930s, had been slated for demolition, but a group of local advocates headed by now-deceased Bill Decker lobbied for its preservation," according to Tim O'Donnell, Manager of Smith Field in 2009.[44] With its three large rolling doors built by Truscon Steel Company, it was among the few hangars in the nation that could easily accommodate the ninety-five-foot wingspan of the DC-3, a twenty-two-passenger twin-propeller plane introduced in 1936. During World War II, the Aviation Packaging Company worked behind the doors of Hangar Two, secretly assembling Interstate Aircraft TDR-I assault drones. The unmanned twin-engine aircraft carrying two-thousand-pound bombs could be directed to enemy targets—an early version of the cruise missile. The TDR was only recently declassified, bringing Smith Field's formerly secret history to light. Conveniently, the drones could be "put on railroad cars right behind the hangar," according to Roger Myers, curator of the Greater Fort Wayne Aviation Museum.[45] Also during WW II, wing covers and tow targets were made for the Army Air Corps on the second floor of Hangar Two, Myers reports.[46]

European and the Asian war fronts. He had left Fort Wayne at a time when the ground was literally being laid for a positive and permanent development in the city's and the region's aviation history, but at a difficult time for his hometown's civil air industry. TWA had added two more daily east-west flights, and the city was negotiating for more commercial flights.[47] Smith Field was barely managing to function with its increasing burdens. With Cornish leaving, Fort Wayne would now have to address the plight of the recently renamed Smith Field without the help of its longtime chief air advocate. He would devote his expertise to doing everything within his power to help the war effort.

8

Keeping the Home Skies Safe

In May 1941, Cornish returned to Fort Benjamin Harrison, where he had often attended annual reserve training, but this time he was there for a physical exam. If found to be fit, he would be required, "without delay," to report for duty to Barksdale Field in Shreveport, Louisiana, with the Air Corps Advanced Flying School, where bomber crews trained for war. Having passed his physical, Cornish was reactivated, but he cheekily requested assignment with the 46th Air Base Group, to be located at the new Baer Field in Fort Wayne. "Not favorably considered" was the military's response—and in July, Captain Cornish was on his way to Louisiana. He left with the blessing of Fort Wayne's Board of Aviation Commissioners and a "leave of absence for the duration of the emergency," and with assurance from the military that he would "be relieved from duty in time to enable him to arrive at his home on 7 July 1942, on which date he will revert to inactive status."[1] That was a promise the military did not honor, as it turned out that he would be needed for the duration.

Barksdale Field, where Cornish was headed, had risen from cotton fields in the early 1930s.[2] Home to pursuit and attack groups, its immense acreage was ideal for honing gunnery and bombing skills. While there, Cornish was executive officer of the 6th Air Base Group; he soon found that he enjoyed working with the ground crews who were supporting the simulated European combat operations. He first flew a Douglas B-18M powered by two Wright R-1820 radial piston engines and then an AT-6A (North American Aviation T-6 Texan), a blunt-nosed,

While stationed at Barksdale Field in 1941, Cornish flew a Douglas B-18M shown at left (author's collection) and a North American AT-6A Texan similar to the ones shown below (U.S. Air Force photo).

shiny metal advanced trainer with a single prop powered by a Pratt & Whitney R-1340-49 Wasp radial engine. Flying those aircraft must have been a thrill for him.

But Cornish's time at Barksdale Field would be brief. In October he was abruptly relieved of the assignment in Louisiana to begin a new one in Washington, D.C. It was a call from an old acquaintance, Major A. B. McMullen, that initiated this second move in four months for Cornish and his family. McMullen and Cornish had first met in 1938, when "Mac" was with the Civil Aviation Administration and "Cap" was running things at Fort Wayne's primary airfield.[3] Their paths crossed again when McMullen and the media went to Barksdale Field in October to observe a bombing demonstration put on by members of the U.S. Army Air Forces (AAF), Navy, and Marines. Two days later, McMullen, who served as the AAF's representative on the Interdepartmental Air Traffic Control Board (IATCB), called Cornish. McMullen needed him to serve as chief of the Flight Operations Division, Air Staff, putting Cornish in a position to act as McMullen's alternate on the IATCB, which also included representatives from the Navy, the Civil

The Cornish family's Ford sedan is parked in front of their residence at Barksdale Field in 1941. (Author's collection)

Newly recruited to serve with the U.S. AAF in WW II, Cornish wears his full military uniform in this official portrait.

Aeronautics Administration (CAA), and the Civil Aeronautics Board (CAB).

In volume 6 of their seven-volume history *The Army Air Forces in World War II*, Wesley Frank Craven and James Lea Cate cite two reasons for the IATCB's formation—complaints about "operational hazards caused by unskilled civilian flying" and "the growing problems of congested air traffic."[4] The board's mandate was to oversee and set rules for civilian and military air traffic and to control and coordinate the use of navigable airspace over the continental United States. Eventually all decisions about site locations for military air installations would be funneled through the IATCB.[5] But despite the interesting aspects of this and other responsibilities offered to him by McMullen, "I didn't like any part of it," Cornish would say later of this move to Washington.[6] But at least one associate, Irv Baldwin from Fort Wayne, thought that he was the ideal person for the task. As he wrote to Cap[7]: "I can see that your new work, along with Major McMullen, must be extremely interesting, and work for which only a very few people in the country are qualified. I mean that direct experience in both military and commercial aviation is required. There are plenty of individuals who have a wealth of experience in one field or the other, but there are not many like yourself who understand both." Putting his personal wishes aside, Cornish assumed his new position around the time of his fifty-third birthday, November 10, 1941.

The United States was unprepared for the horrific event the next month when on December 7, Japan bombed U.S. naval ships at Pearl Harbor. The tasks ahead were monumental. While Japan and Germany had begun to build up their war personnel and materials years earlier, in 1931 and 1933, respectively, the United States had remained an outside observer, maintaining a comfortable distance from the hostilities while providing material support to others who were beginning to suffer as Hitler advanced through Europe while Japan ravaged portions of China. With astonishing swiftness, the United States was at war on two fronts. The nation would have to move quickly to overcome the lead of its enemies and implement a significant buildup of its armed forces. Within a year, the Army alone grew to more than 4 million officers and men, with a planned strength of 7.5 million by the end of 1943.

Lieutenant Cornish, Lois, and their only child, Ruth Ann, enjoy the cherry blossoms a couple of years after their move to Washington, D.C. (Author's collection)

Men and women from myriad occupations willingly answered the call to serve their country in the military. Hundreds of thousands were recruited to work behind the scenes in administrative positions to support the combat troops and to protect the homeland. Cornish was among the latter. He made the transition from managing a busy municipal airport in the Midwest, to coordinating pursuit and bombing practice in the South, to protecting U.S. citizens from coast to coast against enemy air attacks and internal disasters.

History understandably gives maximum attention to the drama of combat, where men and women literally risk life and limb as well as sanity to defend their country. However, attacks on the United States from enemy forces were real, and they would have been far worse had it not been for President Franklin D.

The Interdepartmental Air Traffic Control Board met for the first time on April 7, 1941, and was abolished on May 31, 1946.[8] The minutes from April 11, 1942, state the following:

> The primary duty of the Board shall be the coordination and adjustment of the various activities conducted in the navigable air space above the continental United States, and such other areas as may be designated by the Secretaries of War, Navy and Commerce and the Chairman of the Civil Aeronautics Board, in order to assure the efficient and safe operation of military and civil air traffic.
>
> In obtaining the above object, the Board shall consider the interests and requirements of military, commercial, civil and private establishments and individuals.
>
> The Board shall conduct or direct such studies, research and investigations of the present or proposed uses of the navigable air space as may be necessary in the performance of its duties.
>
> The Board shall recommend such standards, rules, regulations and restrictions as may be considered necessary to obtain the coordinated, efficient and safe operation of military and civil air traffic and the use of the navigable air space by other activities.
>
> The board shall submit such recommendations as it may find necessary to appropriate agency.
>
> All proposed changes in existing policies, regulations, etc., affecting the use of airways,[9] airports, air navigation aids, serial bombing and machine gun training areas, flying schools and other activities taking place in the air space normally or occasionally utilized by aircraft shall be submitted to the Board for comment and recommendations before being adopted.
>
> The general plans for the construction, development, location or relocation of all airways, airports, air navigation aids, aerial bombing and machine gun training areas, military and civil flying schools, barrage balloon, search light training areas, etc., and all other facilities or activities that may effect the navigable air space, shall be submitted to the Board for review before construction or use is authorized, in order that activities of the various governmental agencies concerned may be more closely coordinated, and overlapping or interference of activities avoided insofar as possible.
>
> The members of the Board, as representatives of the respective Secretaries of War, Navy, and Commerce and Chairman of the Civil Aeronautics Board are authorized to secure from their respective departments such assistance as may be necessary from time to time to process efficiently and expeditiously matters referred to the Board.

Roosevelt and the military forces (and civilians) charged with keeping the skies and land safe from hostile infiltration. Besides conducting operations outside the United States and its overseas possessions, the AAF was charged with preparing for and executing air operations in defense of the continental United States. Cornish was a part of that security team, one that took some time to get up to speed.

The nation's lack of preparation is well described by Craven and Cate. "The United States in 1940 possessed only a few elements essential to air defense; it had neither a system nor a single agency responsible for protection against air attacks. By 7 December 1941 the AAF emerged as the military agency responsible for that defense." The AAF developed "a system of unified air defense for cities, vital industrial areas, continental bases, and armies in the field" and "was concerned only with the problem of protecting important areas and installations by interception and destruction of attacking enemy forces," according to Craven and Cate. "The success of the attack on Hawaii had put the United States on the defensive and suggested that the homeland itself was not beyond the range of similar carrier-borne air assaults. For the first time since the War of 1812, the country faced a serious threat of attack."[10]

Oregon was the only one of the forty-eight states to suffer civilian casualties during the war. On a May day in 1945 near the town of Bly, a strange object caught the attention of Elsie Mitchell, a church pastor's pregnant wife, and five children as they were enjoying a Sunday school picnic. When they went up to see what it was, it exploded, killing them all. What they had found was a balloon bomb; launched from Japan, it had floated across the Pacific on a jet stream. The Japanese hoped that once these hydrogen-filled paper balloons reached the United States, the explosive devices they carried would ignite, causing massive forest fires that would kill civilians. Japanese records indicate that approximately 9,000 balloon bombs were launched; it is believed that approximately 1,000 made it all the way to North America, although fewer than 300 have actually been documented.[11] Of course, there were other war-related deaths within the United States. Thousands of AAF pilots, personnel, and airplanes were lost between December 1941 and August 1945 in more than 52,000 accidents. An estimated 1,700 deaths of non-military U.S. citizens were also attributed to the war and occurred abroad.[12]

In the early months of the war, both U.S. coasts had inadequate warning systems and were truly vulnerable. German and Japanese submarines sank merchant ships along the Atlantic and Pacific coasts, respectively; the Japanese submarine

I-17 bombarded the Ellwood Oil Field north of Santa Barbara, California, although fortunately it did little damage. Things improved with a developing radar network, thousands of volunteer aircraft spotters (including members of the Civil Air Patrol), civilians at thousands of ground observation posts, air raid sirens, barrage balloons, and blackouts. But citizens on both ends of the country were on high alert. In Oregon, for example, "a wartime mentality set in immediately. Coastal residents were commanded to black out their homes nightly, covering windows with shades and blankets. 'Block wardens' patrolled neighborhoods, looking for telltale lights and reprimanding offenders. Volunteers watched for airplanes, soldiers patrolled the offshore waters along with the Coast Guard, and coast watchers and their dogs walked the beaches, looking out to sea."[13]

Clearly, defense of the homeland was critical, not only from the standpoint of enemy infiltration, but also because of the conflicting uses of airspace during this frantic time. The tasks that were laid upon the IATCB weighed heavily. Decisions they made often had life-or-death consequences. As chief of the Flight Operations Division and in his other assignments, Cornish helped the board establish regulations for the control of both military and civilian air traffic and formulated plans for the War Department to control civil aviation systems. It was a complicated but interesting job that required him to deal with all branches of the War Department, the Navy Bureau of Aeronautics, and the Defense Commands. His team investigated locations for airports, defense plants, gunnery ranges, and other establishments with the intention of maximizing the usefulness of airways; coordinated military, naval, and civil air traffic on those airways; and cooperated with the CAB, the CAA, and the airlines regarding airspace and with the Weather Bureau on meteorological matters. "We are in someone's hair most of the time," he wrote to his longtime friend Ross Parnin, "leaving us in a mental condition worse than physical fatigue."[14]

The minutes from the IATCB meeting on September 9, 1941, illustrate the nature of the issues the board grappled with, many of which revolved around airspace reservations and danger areas. Airspace reservations, similar to today's no-fly zones, are areas over which aircraft are prohibited by order of the federal or a state government. As of mid-July 1942, two such areas had been established by executive order—Hyde Park, New York, the location of President Roosevelt's home; and the District of Columbia.[15] FDR was opposed to adding more airspace reservations than were absolutely necessary because of their effect on aerial navigation. Danger areas, on the other hand, evolved into something far more complex as the war unfolded. The initial authority for designating

Craven and Cate write: "In that headquarters there was throughout the war a willingness to experiment with administrative procedures and agencies; offices changed designations with a rapidity that was reflected in almost each new Pentagon telephone directory, and in some instances the transformation may have seemed hardly more than finding new titles for familiar officers performing familiar tasks in familiar rooms. In part the experimental mood derived from the lack of any deeply entrenched traditions, in part from the liberal use by Arnold of Civilian personnel and techniques in office management."[16] Perhaps that partially explains the many titles and assignments Cornish held in the military in Washington during World War II, as listed in his "AAF Officers' Qualification Record":

10/20/41–3/9/42
    Asst. Chief, Civil Aviation [also written Aeronautics] Section, Air Staff, Headquarters AAF

3/9/42–9/15/42
    Asst. Chief, Military Air Traffic Control Division, AFOCA

9/15/42–1/9/43
    Director, Military Air Traffic Control Division, War Organization & Movement, Headquarters AAF

1/9/43–5/29/43
    Chief, Information Division, Director of Flight Control, Headquarters AAF

5/29/43–10/10/43
    Flight Control Branch, Technical Service Division, Headquarters AAF
    AC/AS Operations, Com. & Reg. Washington DC

10/10/43–7/1/44
    Chief, Operational Services Branch, OFS, Office Deputy Chief, Washington DC

7/1/44–duration of service
    Chief, Flight Services Branch of the Flight Operations, Headquarters AAF, AC/AS, Training Flight Operations Division Responsible for Provisions of AAF Reg. 63-1 (Hazards in Air Navigation) coordinating with Navy, CAA and AGS Administration; Advisor on instrument approval procedures
    Served throughout as Alternate, War Department Member, IATCB

danger areas rested with the administrator of civil aeronautics, and they were intended simply to protect personnel on the ground, not aircraft in flight, to reduce the risk from aircraft making forced landings, dropping objects, or accidentally falling on sites such as ammunition depots, ordnance plants, fortifications, and arsenals. However, once the United States had been drawn into the war, the IATCB recognized the potential danger that sabotage and subversive activities posed to high-explosive areas, such as arsenals, ammunition depots, shell-loading plants, torpedo stations, and fuel depots. Before the war, it had made sense to highlight explosive areas on aeronautical charts. A flaming bomb symbol was used to mark such locations on maps, while "Hi-X" signs and distinctive lights bordered the actual danger zones, making it easy for airmen to identify and avoid those areas.

On August 1, 1941, however, aviators had been advised that the signs would be removed from Navy ammunition depots and that all lights were subject to being temporarily extinguished without prior notification, as these could "provide valuable information to subversive elements and saboteurs."[17] Similarly, the War and Navy Departments requested that flaming bomb symbols be removed from all aeronautical charts to prevent the enemy from using them to pinpoint vulnerable targets. These changes made it tougher for a potential enemy to identify targets, but they also posed a problem for U.S. forces, who could no longer easily identify all danger zones. Meeting minutes show that the IATCB agreed unanimously that "only areas in which it was hazardous to navigate aircraft should now be considered or classified as Danger Areas" and that "aircraft shall not fly lower than 1,000 feet except when taking off or landing, and then in such a manner as to permit an emergency landing outside of such areas in the event of power failure." Examples of designated danger areas were Fort Belvoir and the Dahlgren Naval Proving Ground near Washington, D.C. The board agreed that the War Department should follow the Navy in removing ground identification markers and true lights from the boundaries of high-explosive areas.[18]

Additional minutes from the September 9 meeting attest to the complexity of the problem. Item 19 reads: "There is a decided lack of coordination within and between the various branches of the Army and the Navy, and between the War, Navy and Commerce Departments, with respect to the establishment of artillery firing ranges, aerial bombing and gunnery ranges, search light and barrage balloon areas, military and civil flight training schools, high explosive areas, etc., and no uniform procedure and policy with respect to the filing of applications and plans for, and the establishment of, danger areas or air space reservations."

Years later, Cornish recalled one example of the tragic ramifications of unclear directives: "Industries that built goods requiring high security put machine guns on their roofs and the operators were told to shoot down anything that flew over. Two people in a small private aircraft flew over the Winchester plant [the St. Louis Ordnance Plant] and were shot down."[19] It was a case of misidentification and panic on the part of the machine gun operators, he concluded.[20]

The problem extended well beyond manufacturers of armaments. Item 20 of September's minutes reads: "This lack of coordination has resulted in the establishment of Army and Navy training areas which overlap each other; and AAA [antiaircraft artillery] firing areas, and aerial bombing and gunnery ranges being established within long established civil airways." In the future, the board stated, "all applications for the establishment of Air Space Reservations or Danger Areas [will] be routed to the appropriate member of the Interdepartmental Air Traffic Control Board for review, coordination and recommendation by the Board."

Cornish arrived two months later, officially reporting for duty on November 11. The following day, he sat down in an office at Gravelly Point with military and civilian officials for his first IATCB meeting, chaired by McMullen. After leaving Barksdale Field, where he had found his work with bomber crews to be immensely satisfying, he was entering an understandably confused arena, given the newness of the nation's danger alert. After listening to the discussion of various issues and positions, Cornish perceived the difficulty of finding workable solutions to problems that existed between divergent parts of government. They sometimes worked at cross-purposes in the rush to build forces and munitions during those months leading up to the formal declarations of war in late 1941—on December 8 against Japan and on December 11 against Germany and Italy. For example, special trench mortar powder was manufactured and tested at the Radford Ordnance Works and Proving Grounds near Radford, Virginia, within the horseshoe bend of the New River at White Horne. Trench mortar projectiles on that range could reach a maximum height of approximately six thousand feet, and firing was conducted daily. At the time they did no night firing, but if production increased, that could change. The problem? Green Airway No. 4, which connected Roanoke, Virginia, and Knoxville, Tennessee, passed directly over the proving grounds; the testing of projectiles threatened safe air traffic between the two cities. The CAA contended that because the airway followed a natural valley flanked by a mountainous landscape, rerouting aircraft would force them to fly over rugged and potentially dangerous terrain. In addition, there were no funds available to relocate the lights, radio range stations, and intermediate landing fields that served the airway. It was pointed out that neither the CAA nor the IATCB had been advised before the development of the Radford facility. The board recommended that since the activities on the proving grounds constituted a serious hazard to air traffic on this airway, no firing into low clouds or in overcast conditions would be permitted, and the grounds would need to be relocated outside the confines of Green Airway No. 4. However, the board then backtracked somewhat, having decided to allow firing pending the establishment of a new proving grounds. Until then, safety officers stationed in strategic locations would observe air traffic and communicate directly with control officers by "interphone" when it was necessary to hold fire because aircraft were within the danger area.

Clearly, bombing and testing and artillery ranges posed a danger to aircraft within both civil and military flight paths. Surprisingly, the minutes show that civil air traffic often took precedence in the board's final recommendations. Cornish would soon become familiar with the IATCB's decisions in this regard. He would also discover that no part of the country was immune from conflict between those involved in vital defense activities on the ground and those needing to use the airspace above.[21] For example, the Arlington Bombing Range, between Pendleton and Portland, Oregon, fell in the center of Red Airway No. 1; the Southwestern Proving Grounds were under Green Airway No. 4 between Texarkana, Texas, and Little Rock, Arkansas; the Sheep Mountain Gunnery Range was almost entirely within Amber Airway No. 2 between Las Vegas, Nevada, and Salt Lake City, Utah; the Fort Knox Armored Forces Range in Kentucky fell within the confines of Amber Airway No. 6 between Louisville and Nashville, Tennessee.[22] Additionally, the Camp Callan Artillery Firing Range was located almost entirely within the confines of Amber Airway No. 1 between San Diego and Long Beach, California, a course used extensively by both civil and military aircraft. The IATCB minutes noted that "training at Camp Callan includes firing of small arms, antiaircraft machine guns, automatic weapons, and 3-inch guns, and 155-mm guns, and the use of antiaircraft searchlights; that searchlights and aircraft without lights may be operated within the limits of this range in connection with training at night; that trajectories of projectiles from A.A. [anti-artillery] guns fired on this range may reach an altitude of 30,000 feet; that firing on this range may be conducted both day and night from 8:00 a.m. Monday until 12:00 noon, Saturday; holidays excepted." The board made several recommendations, including "that the activities at this Camp be transferred to other locations as rapidly as possible."[23]

Wartime initiatives and resulting activities across the country moved at breakneck speed, and it was important for anyone who entered the airspace to stay informed. Flight maps for pilots indicated areas that required avoidance or special attention, including "Prohibited Areas (Air Space Reservations)"; "Caution Areas . . . in which . . . there are visible hazards" such as radio towers, masts, flagpoles, smokestacks, or water towers; and "Danger Areas . . . in which there is an invisible hazard such as might exist over an artillery or bombing range." During the war, aerial bombing and gunnery ranges and areas being used for flight training, artillery practice, and searchlight training were all classified as danger areas to be avoided by other aircraft.[24] And according to the board, areas where barrage balloons were in use constituted "the greatest of all present hazards to aircraft in flight." At the IATCB meeting on December 5, 1941—two days before the bombing of Pearl Harbor—one of the topics of discussion was how best to mark barrage balloons so as to warn pilots away from them and thereby prevent fatal accidents.[25]

Having been used by the British during World War I to defend against a low-level air attack by the enemy, barrage balloons—shaped like fat footballs with fins—were redeployed over Britain by the thousands in World War II. The first American balloon was constructed and sent aloft in February 1942 at the Army's new Barrage Balloon Training Center at Camp Tyson in Paris, Tennessee. It was hoped that barrage balloons would be an effective deterrent against an enemy invasion. With the use of a motorized winch, they could be strategically raised and lowered, and they could be moved from one place to another as needed.

The U.S. military had high hopes for the use of barrage balloons in deterring an enemy attack on the homeland, but the "fat footballs with fins" created difficult issues that Cornish and the IATCB had to resolve. (Photo by Alfred Palmer or Pat Terry, Parris Island, South Carolina, May 1942, LC-USW3-002371-E, Library of Congress Prints and Photographs Division, Washington, D.C.)

They were often grouped into a circle or used to form a "barrage battalion." Flying into a barrage balloon or one of the heavy steel cables that tethered it to the ground was extremely hazardous—and not only for enemy aircraft.[26] Friendly aircraft thus needed to be properly warned. But how? A single battalion might have fifty-four balloons operating in a four-square-mile area. Marking them all would not be easy.

In light of the buildup of this defensive program, the concern about warning markers was well-founded, and Cornish was asked to present ideas to the IATCB. He listed several possibilities, including positioning red rotating beacons at the corners of the area, having "beacons oscillating in a vertical plane of 180 degrees as an alternate," using two radio ranges with the "legs" set "so as to box the area," and "attaching streamers and warning lights on cables at the outer edge only." By April, approximately 430 barrage balloons were in place along the U.S. coasts, protecting vital cities, factories, and harbors from invasion by German forces on the East Coast and Japanese forces on the West. Unable to reach a definitive conclusion regarding the problem, the IATCB recommended that "no attempt be made to promulgate general regulations governing the operation of barrage balloons in defense areas as tactical and strategical situations will determine when, how, and at what altitudes these balloons will be flown."[27] The matter was passed along to the CAA's Technical Development Division for further study. At the very least, barrage balloon locations would be flagged as danger areas on aeronautical charts.

The inability to make a timely determination about important matters hampered the functioning of the IATCB throughout the early days of the war. Deliberation could be cumbersome, and crucial decisions were often delayed. Different divisions and individuals in the agencies had to weigh in; papers had to be written, positions stated and debated, and territories staked out. Before taking action and making recommendations on dilemmas such as whether to mark barrage balloons with lights, staff members had to develop position papers for consideration by the IATCB. Recommendations were then taken back to various military and civilian departments in the Army, Navy, CAA, and CAB. The steps necessary for effecting change among the various bureaucracies could be frustrating and had the potential to be dangerously time-consuming.

According to a document found in one of Cornish's AAF binders, the War Department's job was made all the harder by the poor preparation of essential materials. In a general directive to subordinates dated September 17, 1942, Major General George E. Stratemeyer made it clear that the chief of the Air

Staff needed to be protected from "half-baked ideas, voluminous memoranda and immature oral presentations." Staff were to do their job properly, so that all he would have to do was indicate his approval or disapproval of the suggested action.

> The more difficult the problem the more the tendency is to present it to the chief in piece-meal fashion. It is your duty as a staff officer to work out the details. You should not consult your chief in the determination of those details, no matter how perplexing they may be. . . . It is so easy to ask the chief what to do, and it appears so easy for him to answer. Resist that impulse. You will succumb to it only if you do not know your job. It is your job to advise your chief what he ought to do, not to ask him what you ought to do. He needs answers, not questions. Your job is to study, write, restudy, and rewrite until you have evolved a single proposed action—the best one of all you have considered.
>
> The rough draft must not be a half-baked idea. It must be complete in every respect except that it lacks the requisite number of copies and need not be neat. But a rough draft must not be used as an excuse for shifting to the chief the burden of formulating the action. Avoid submittal of hastily prepared inaccurate material lacking a concise, specific, workable recommendation.
>
> Finally, when you have finished your "completed staff work," the final test is this: If you were the chief would you be willing to sign the paper you have prepared, and stake your professional reputation on its being right?[28]

Considering Cornish's lifelong attention to detail, he would surely have heartily agreed with Stratemeyer.

There was no confusion about the meaning of the following safety bulletin, which was prepared by the Civil Aviation Board and included in the IATCB minutes from July 1942, with a note that, "danger areas" should be substituted for "restricted areas":

Don't Get Shot Down

In these days of war, failure to read the airport bulletin board daily may prove to be fatal. New restricted areas are being established from time to time and these areas are likely to be protected by anti-aircraft batteries and machine guns. The orders to the gun crews are to shoot down any civil aircraft flying over those areas, and these orders are likely to be carried out with dispatch. There has already been a case in which a pilot did not read the airport bulletin board notice telling of a restricted area. He carried sight-seeing passengers over this area, was shot at, and one of his passengers seriously wounded.

Reading the bulletin boards for these notices does not mean looking at them once a week, but before each flight the pilot should make sure that no notice has been added which he has not seen. Never lose sight of the fact that we are at war and our armed forces are not fooling when they set out to protect a defense area. They mean business!

Life may have been going on pretty much as normal back home in Fort Wayne, but from his vantage point in the nation's capital, Cornish, who had been promoted to major on February 26, 1942, and then to Lieutenant Colonel on December 5, 1942, was well aware of the very real dangers that the country was facing.

9

Calming the Turbulence

THROUGHOUT THE WAR YEARS, mail from home kept Cornish apprised of the woes back at what was now Smith Field, most resulting from the damage that the heavy Army transports were doing to the runways, which had yet to be upgraded.[1] In 1942, TWA threatened to pull out of the facility if the runways continued to deteriorate. The War Department had not followed through on the promise to provide facilities for commercial traffic at the new Baer Field; nor had it remedied the steadily worsening conditions at the smaller airport belonging to the citizens of Fort Wayne. By the spring of 1944, Irv Baldwin wrote to Cornish:

> Fort Wayne Aviation is just about eliminated. Most of the airplanes belonging to people who have supported and built Fort Wayne Aviation have been moved to other fields. I have taken one of my ships to Indianapolis and the other will be flown out of the mud-hole at Smith Field as soon as the major overhaul on the engine has been completed.

Adding insult to injury, Baldwin continued:

> Renting of the Smith Field hangar was not with the Navy as thought but to the Aviation Packaging Company. Very suddenly one day, in the midst of the blizzard we had here, and in the coldest weather of the winter, all airplanes were moved out of the hangar into the snow banks and tied down. This, of course, pleased everyone except those who owned the

airplanes. We were told not to get excited, that this was a patriotic situation at Smith Field, that we might as well agree to the situation because if we didn't, the government would take over the field in its entirety. We were told that a new hangar would be built by the city immediately and certainly within sixty days the new hangar would be far enough along to provide shelter for the airplanes belonging to the citizens of FW (the new hangar would, of course, be built at the expense of FW citizens).... Two lines of airplanes are being subjected to the full blast of the spring winds, rains, snows, and sleet and yet no hangar has been started. Bids were received but no contract placed.

The citizens of Fort Wayne were understandably frustrated and disappointed.

Baldwin sent Cornish additional heated letters. On October 11, 1944, for example, he wrote: "Regarding airline service, the runways at Smith Field have disintegrated where they are no longer safe for us. They are full of ruts, and when a heavy ship lands, chunks of asphalt are thrown in all directions. The airliners, therefore, have to use the turf instead of the runways, and when the turf is at all wet, the airliners skip Fort Wayne. No night landings are permitted at Smith Field by TWA. Don't you think this is a fine state of affairs for an important industrial center like this?" Cornish recommended that Mayor Baals write directly to Lieutenant General Henry H. "Hap" Arnold, "as the existing lease by and between the City of Fort Wayne and the War Department is still in full effect and should be the basis for any claims for repairs or replacement of the damaged runways on Smith Field." This course of action was followed, but as of the spring of 1945, there was still no facility at the newer Baer Field to handle the heavy traffic of scheduled airline flights. The war effort had grabbed the greatest attention and left the hinterland to suffer.

Still, even as the city's sons were fighting overseas and Cornish was busy sorting out civilian and military airspace in Washington, some of the updates that he received from home were of a lighter nature. Ed Wrib, who was employed at Smith Field, wrote in 1941: "A couple of weeks ago Pete Anders placed a slot machine in the restaurant and I told him he had to remove it as the City Municipal Airport was no place for such a machine. He did not feel so good about it, but he took it out just the same and we are again friends." And this report came from Cornish's Fort Wayne associate Fred Romy in the winter of 1942: "There isn't much change in the airport, except we were without a restaurant for about a month. Phil Greeley and his wife have taken it over now, and we have beans once again."

Friends in the civilian world looked to Cornish, with his experience, connections, and proximity to policymakers, for help with issues related to ground and air. In other words, they wanted special treatment arranged by their insider friend. Some hoped to take advantage of the accelerating need for pilots by providing classes as part of the Civilian Pilot Training Program. On February 25, 1942, Fred Romy, president of the Inter-City Flying Service at Smith Field, wrote that he would like to do his bit in "Keeping 'EM Flying" and wanted to see "quite a few Stearman trainers on the line here [in Fort Wayne]."[2] Cornish replied that all such activities must be secondary to the objective of winning the war, but he advised Romy to initiate action to secure a primary training contract and, presumably, procure the desired Stearmans. Things worked out in Romy's favor, as he shared in an August 1943 letter: "Things are rolling along here fairly well. We now have our airplanes released by the War Production Board for approved private flying and have about sixty approved students now flying with us. Again, thanks for everything."

In the spring of 1942, Cornish received a letter from A. R. Stimson, the city attorney in Huntingburg, Indiana, who wrote that his community had the facilities for a civilian or military training program and wanted help in getting things started.[3] "I will be indebted to you for life, but if you cannot do this yourself, make some suggestions to me and pull a few strings there in order to bring it about." After Cornish sent a less than encouraging response and suggested that the airport be developed as an emergency field, Stimson complained, "What I cannot quite understand is why you will not be in a position to give me a little help in some form of army aeronautics development." This time Cornish softened his reply. The War Department was besieged by congressmen and citizens

> As aviation historian Janet Bednarek writes, "During World War II . . . both general aviation pilots and manufacturers found ways to participate in the war effort. Pilots organized the Civil Air Patrol, an organization that eventually became an auxiliary of the Army Air Forces (and later the U. S. Air Force). Civil Air Patrol pilots performed a number of duties. They flew coastal patrol missions looking for enemy submarines while others flew over the nation's forests acting as fire spotters or flew humanitarian missions such as emergency medical flights and supply drops to areas hit hard by blizzards, floods, or other natural disasters. Their activities also helped keep a large number of general aviation airports open and active during the war."[4]

with special requests, he reminded Stimson, and any request would have to fit the needs and plans of one of the military branches. He advised Stimson not to plan on Huntingburg's facility being selected but, given the tremendous expansion of both War and Navy Department programs, his request "might be selected for some type of facility." In closing, Cornish wrote: "Keep working and surely your efforts will be rewarded."

Greenville Aviation in Ocala, Florida, was headed by a Cornish acquaintance, Frank A. Hanley, formerly from Muncie, Indiana, and K. A. Spurgeon. It began providing primary training for Army Air Corps cadets in October 1941 at Jim Taylor Field (now Ocala International Airport).[5] In a letter to Cornish in the spring of 1942, Hanley wrote that Brigadier General Donald H. Connolly, the administrator of civil aeronautics in the Department of Commerce, was "casting covetous eyes at the one and only airplane that we own, namely, a Stinson Reliant," which presumably was being used for private purposes. Hanley understandably wanted to keep the plane. "Will you do a little pitching for an old friend?" he implored. Cornish offered little consolation over the very real possibility that Hanley would have to sell his aircraft to the Army, explaining that the War Department wanted all such aircraft that were "not necessary to the owner or not being used in activities that were vital to the war effort."[6] If the school had been a Civilian Pilot Training contractor, or if Hanley had had another school or two and was required to travel quickly between them for administrative or other official purposes, it would have been a comparatively simple matter to arrange a deferment so that he could retain the Stinson. After checking, however, Cornish determined that Hanley's situation did not qualify. "We are at War and you know what that means to us, as it affects our individual rights and conveniences," he concluded. Cornish held clear-cut views of how things should be, with the war effort paramount.

Such matters briefly diverted Cornish's attention from what he was in Washington for: keeping the skies safe. When he arrived in Washington in the fall of 1941, the Army Air Forces offices were housed in some single-story frame buildings at Gravelly Point, near the new Washington National Airport. Construction of the Pentagon had just gotten underway. As each section was completed, its occupants moved in. Seventeen months after ground was broken, the world's largest office building, designed to accommodate forty thousand employees, was complete. In mid-August 1944, Cornish wrote to his friend Ross Parnin, "It certainly is great to work in an air-conditioned office building, particularly during these hot days."[7]

During Cornish's nearly four years in Washington from 1941 to 1945, a steady stream of issues involving intercontinental air traffic required his personal attention. He addressed complicated, sometimes confusing, and occasionally dramatic matters that involved different branches of the military and sometimes civilians. In a 1992 interview, he discussed the variety of concerns that his office faced and shared stories that connected him to Indiana's flatlands, the World Series, Black Widows, gliders, and Asian pilots.[8] He also talked about the role he played in the establishment of an important Indiana air base, Bunker Hill Naval Air Station, located on what had been farmland between Kokomo and Peru. "Did you ever wonder how these 3,140 acres of prime, landlocked Indiana farmland were selected as a base to train Navy, Marine, even Canadian and British air cadets to become the pilots of World War II?" asked *Mission Briefings* in its September 1991 issue. "The answer is simple, even ironic. A Hoosier in Washington, proud of his heritage, was disturbed to see that Indiana was being left out of the almost panic military base building as the U.S. prepared for war."[9] During the interview, Cornish recalled:

> Federal airways were highways in the sky and needed protection from ground activity, but those airways became crowded. Technically you could not fire an automatic .45 in the air without our [Interdepartmental Air Traffic Control Board (IATCB)] approval. I had a map of the continental United States with aeronautical charts. Every military installation in the United States was on that map on the wall. The federal airways were twenty-miles wide on aeronautical charts. We had to reduce the width to 10 miles because of congestion. It got so chock full of activity south of the Mason Dixon line that it was saturated. I was called at home one Sunday morning to come down to the office at 2:00 PM. The Secretary of Commerce wanted to see a map they understood I had. There was a need to go north of the Mason Dixon line for new installations. Navy personnel also wanted to look at the map for a place in the Midwest for building some installations. I said that if they were really looking for a place that's wide open for primary training, I knew a place where they might take a look. Out there in Indiana is an area near Kokomo off the federal airways so that from that standpoint you'd have no problem. The land is level as a table top. Ten days later, one returned. "We'd like that. If we submit a request for approval with the Board, would you approve it?" "Well, I think we'd approve it."

The base opened on July 1, 1942, and more than seven thousand American and British pilots, including the famous baseball player Ted Williams, took their World War II flight training there. Now Grissom Air Reserve Base, it is the largest employer in Miami County.

A later incident, on October 5, 1943, shows the balance that Cornish had to strike in trying to help regulate air traffic while maintaining appropriate wartime demeanor:

> Some politician had written General Hap Arnold about somebody buzzing New York Stadium during the World Series with aircraft. The letter landed on my desk with a red tag on it. They had the number of the aircraft. In a couple of hours we knew where the airplane was from and who had been flying it. I also learned that they had orders to proceed for combat duty. Ultimately, I just dropped that God-damned file. They were headed for worse trouble than for buzzing the Yankee Stadium. All I did was send a letter to General Arnold to forward to the Senator that we regretted the incident very much and assured him that appropriate action had been taken to correct the matter and that we hoped there would be no further difficulty.

Military officials in Bangor, Maine, the flight's next stop en route to England, fined Second Lieutenant Jack Watson and the crew of plane number 98 from the 303rd Bombing Group. Following the crew's well-publicized heroic mission in the European war theater, however, New York City mayor Fiorello H. La Guardia, who had demanded that the pilots be grounded and subjected to court-martial, sent a contrite telegram forgiving them for their low-level circle above the stadium.

Florida's strategic location made it a logical target for alien forces and demanded significant defense measures. Because of its warm climate and abundant flat, undeveloped land, dozens of military installations were established in the state, including numerous Army Air Forces bases for antisubmarine defense in the western Atlantic and the Gulf of Mexico. Not every American is aware of the very real danger the war posed to the waters along the United States' eastern shores. Not long after Pearl Harbor, Germany launched "Operation Drumbeat," using U-boats to attack merchant vessels in the virtually undefended Allied shipping lanes along the East Coast.

German submarines lay off the Atlantic coast watching the American ships go by—fat, slow tankers and freighters. It was a shooting gallery, where the Germans could lie quietly and wait for the most inviting targets. Ships riding high in the water, empty of cargo, were allowed to pass; they were not worth a torpedo. The submarine captains waited instead for tankers wallowing deep in the water under heavy loads of gasoline coming up to the east coast from refineries around the Gulf of Mexico. They were slow, defenseless. . . . So easy that people stood on beaches from Florida to New Jersey and watched in horror as the tankers turned into white-hot torches sinking into the Atlantic. In the mornings, sunbathers would see the bodies of seamen washing ashore in the surf.[10]

At night, the U-boat captains used the light of coastal cities to silhouette their targets. By the time the Navy gained control of the situation, as many as 574 ships had been sunk and nearly five thousand lives had been lost. German spies brazenly came ashore at Ponte Vedra, Florida, near Jacksonville, but were captured before they could blow up the state's railroad lines and halt the shipment of war supplies.

Army Air Force planes and Navy ships from Florida's bases were viewed as an important first line of defense for the Atlantic coast, the Caribbean basin, and the Panama Canal. While some of the state's military establishments were focused on defending U.S. merchant ships and citizens onshore, others were training fighter and bomber pilots and air crews. The number of installations in Florida grew from 8 to 174 during the war, so it is not surprising that the complications of military operations in the air came under Cornish's scrutiny more than once.

He also recounted the secretive operations in Florida involving the Northrop P-61 Black Widow.[11] Named for the venomous spider and painted black to make it harder to see, the specially designed fighter plane was equipped with onboard radar that enabled it to sneak up on aircraft at night. According to an ad for the P-61, "the first designed-for-the-purpose night fighter," the plane had "unusual speed. It can also loaf, almost hover as it hides in the night sky."[12] The P-61's maiden flight occurred in May 1942. That month, the 348th Night Flying Squadron began training at the Fighter Command School, Night Fighter Division, Army Air Forces School of Applied Tactics in Orlando, at what is now Orlando Executive Airport. "With no lights on," Cornish recalled, "they [Black

The use of Northrop's P-61 Black Widow during secretive nighttime training operations in Florida led to a conflict with the Airline Pilots Association. This photo is of a P-61C Black Widow at the National Museum of the United States Air Force, Dayton, Ohio. (U.S. Air Force photo)

Widows] would go up and approach airlines. The airline people got disturbed about that and they complained to the CAB [Civil Aeronautics Board]. The Air Line Pilots Association said they would stop flying down there if it [interception by Black Widows] didn't stop. That got to me. . . . They just wanted to go up there, intercept it and pull away. It was a good operation. We invited a couple of birds, APA people topside, to go down and observe these authorized operations." Once the APA officials understood what was happening, they withdrew their complaints. "Settled that affair," pronounced Cornish.

Gliders had been around since the early years of flight, but during World War II it was decided that they should be enlarged and used to transport personnel and equipment. General Arnold believed in their practicality, and by May 1943, at his urging, the first Waco gliders arrived in England to be assembled. On August 1, however, a glider lost a wing in St. Louis, and all ten passengers, including the city's mayor, fell a thousand feet to their deaths as the spectators watched in horror. Probably a result of faulty workmanship, this accident unnerved the public and shook the faith of some in the military as well. Cornish recalled that when the story of this tragic incident "hit the Headquarters," many presumed that gliders were not safe. When it was suggested that a demonstration might prove otherwise, he recommended that they get a fellow Hoosier, Lieutenant Mike Murphy, to be the lead pilot.[13] "Mike was the kind of pilot who could put an airplane through a maneuver safely where other people would get killed doing it. And I thought, 'Boy, this is the time for Mike to show his stuff.'" A bleacher was built at Fort Bragg's Polk Field for the "top brass." (Although Cornish was on hand, he was not in the "top brass" category.) After a military transport towed Murphy up and released him, "he looped and he spun and he did everything with it then came down and landed right in front of them. Beautiful. And about that time in came a whole flock of gliders and landed out there too. They put on a show. . . . And Mike proved it. They [the military] bought it and went ahead." Thousands of gliders would be built and used during the war; in 2007, a concurrent resolution was adopted in the U.S. House of Representatives to honor "the heroic service and sacrifice of the 6,500 glider pilots of the United States Army Air Forces during World War II."[14]

Following Pearl Harbor, the Air Defense Command was tasked with watching for Japanese planes along the West Coast. One day, in Cornish's words, "things went into a spin" when some voices were heard on radio receivers that the hearers took to be Japanese. But it was a case of mistaken identity. "We were training some Chinese pilots out there," he said.[15] "The airplanes had radio equipment for communications between the student and instructor in the two-place airplanes. They could throw the switch to talk with someone outside or just intercom. They were instructed specifically not to use the radio except for intercom. Well, the word hit that Japs had invaded by air. . . . People out there didn't know Japanese from Chinese. Big alarm." But with clarification came calm.

Cornish later joked that during the war he flew an LSD, a Large Steel Desk. Undeniably, he did spend considerable time putting requests in proper form and attending IATCB meetings, working tirelessly to avoid and arbitrate conflicts in the use of navigable airspace and helping pilots become aware of and learn to identify the hazards of air navigation.[16] He also advised on "instrument approval procedures." When he was not honing his administrative skills, he continued to gain flight experience, including more than thirty hours "under the hood," though presumably using more sophisticated tools and instruments than the ones he had worked with in the 1930s while training novice pilots in Indiana.

The military provided him with training beyond what he already knew about flying in dangerous weather. Among his papers is a thick manual, *Instrument Flying Technique in Weather*, prepared "for personnel rendering service to the United States and its allies." The book deals with life-threatening weather situations, with chapters titled "Weather Elements Affecting Flight," "Turbulence," "Ice Accretion," "Flying Clouds and Fog," "Showers and Squalls," "Thunderstorms," "Cold Fronts," "Warm Fronts," and "The Occlusion."[17] Pilots had the option to avoid flying in such circumstances when it was possible. Cornish had no desire to press his luck when conditions threatened safe travel; after having to cancel his plan to make a quick trip to Fort Wayne to visit family and friends in early May 1943, he wrote to his lifelong friend Ross Parnin, "So inasmuch as I am bound to my conviction of living to be one of the oldest, though not the best, pilots, I decided not to try the trip to Fort Wayne in bad weather."

During the latter part of his wartime tenure, Cornish and Colonel Joseph S. Marriott crisscrossed the country to meet with IATCB regional subcommittee members in New York, Atlanta, Fort Worth, Kansas City, Chicago, Seattle, and Santa Monica. Washington staff flew to each of those cities at least two times a month in aircraft provided by the Air Force, the Navy, and two civilian agencies. The flights enabled Cornish to fulfill the requirements for classification as senior pilot, or first pilot, for multi-engine planes, specifically C-78s, B-18s, AC-45s, and C-60s. To qualify, he had to document ten hours of flight time in each of those aircraft over a twelve-month period. By the end of that period in 1945, he had logged 817:30 hours.

In January 1945, "Cap" heard from A. B. "Mac" McMullen, the old military buddy who had recruited him in 1941, and reported back that the IATCB was still grinding out its findings and recommendations, that there were seven regional committees and several subcommittees, all doing a good job, and that the board met three afternoons a week and was on call at other times. The confusion among

During the latter part of World War II, Cornish, shown here in the cockpit with Colonel Marriott, routinely made cross-country flights to meet with IATCB subcommittees. (Author's collection)

Military and civilian members of the IATCB attended a farewell dinner for Colonel A. B. McMullen on March 16, 1943, as depicted here by a cartoonist. Cornish is fourth from the left; McMullen, sixth. (Author's collection)

the various entities during the early years of the war seemed to have resolved itself somewhat. Mac was by then deputy commander of the North African Division of the Air Transport Command, a region that extended from Dakar to Tripoli and across the Mediterranean into France and Italy. He wrote that he had spent ten enjoyable months in Accra and had had the opportunity to fulfill his boyhood ambition to hunt big game and visit the native tribes in Central Africa before he was placed in North Africa. He revealed that he was living "in one of the finest hotels in North Africa" and concluded that the war was so far "not too tough."[18] Though the war would officially continue for another eight months, the relaxed tenor of this letter written at the beginning of 1945 indicates that militarily, the end was already in sight. Life was about to change for those who had spent their days in combat and conducting behind-the-scenes problem solving. Cornish would soon make the transition out of his military responsibilities.

One of his four nephews was preparing for that same transition. In June 1945, Bob Hockemeyer, the son of Cornish's sister Irma, wrote from Bruck, Germany, asking his uncle's opinion about the educational opportunities being offered to occupational troops. Cornish, whose formal education did not go beyond high school, replied:

> You should take full advantage of every opportunity offered by the Army to more completely round out your education and better prepare you for your return to civil life. Neither you nor anyone else can tell at this time what future advantages will accrue to you by virtue of some previous training or that which you may receive in the Army during your tour of service. You must assume that you will some day return home to resume your normal place in civil life—be it six months or six years from now. So, for your own sake, if for no other reason, take good care of your mind and body so that whatever the period of military service, you will not feel that it was time wasted. On the contrary, you will feel that it was only an interlude during which you gained much in experience and knowledge of the ways of life.

The previous spring, Augusta "Lee" Bowling, who had handled secretarial work for the staff, wrote a memorandum to the IATCB staff, which she titled "Thoughts of Farewell." Dated March 31, 1944, it reads:

> I could look at the wrong side of the picture and often have because there have been Saturday afternoons spent over a typewriter when others were free back in the fall of '41 when the Board was in its infancy. There have been nights of taking notes of meetings long after others were gone. There have been strained eyesight and rumpled disposition. There have been days of disillusionment and days when we all grumbled too much, about one thing or another. There have been days started out tired with too little rest the night before—wishing from 9:00 a.m. for quitting time. There have been days of physical and mental 4F. There have been lulls in the work with nothing better to do than criticize others. And there have been days of poverty before payday.
>
> But all that be as it may, busy days are happy days; and, to offset all of the hardships and disappointments, there have been days of direct contribution to the all-important effort to save human lives; days of dealing with big things far over our heads that we like to feel were conquered in some small degree; days of dealing with great people who are daily shaping

the destiny of a great country; days of disappointment turned bright the next day; and days of feeling within ourselves that we have grown bigger.

We can assume that Cornish felt the same way when he departed for his next challenge back home in Indiana.

Many decades later, Cornish produced a handwritten summary of his wartime involvement:

> World War II
> Ordered to active duty July 5, 1941, at rank of Capt., Barksdale Field, Shreveport, La. Oct. 1941. Ordered to Hdqs Air Service, Wash. D.C. in Air Staff—IATCB. Reported there Nov. 11, 1941. *War declared 12-7-41.* Served as Alternate War Dept. member IATCB (Maj. A. B. McMullen. Made Chief Flight Operations Sec. of the Army Air Staff. Served as such until released from active duty July 15, 1945 to serve as Director, Indiana Aeronautics Comm. Prevented grounding of all civil aviation for duration. (Gen. Spatz) Advocated Federal-civil operation of airport traffic control towers—with military personnel for military traffic in emergency only. Sold Building & Grounds & Navy Bureau of Yards & Docks on principle of new apts. [airports] at locations which would provide for their use as civil airports after the war. Sold Navy on location of primary flite trang. station now Grissom Air Force Base. Drafted Army-Navy Florida Agreement to reduce aviation conflicts in Florida air space. Investigated and solved many air space use conflicts between air & ground and other users.[19]

Joseph B. "Doc" Hartranft, who also served on the IATCB and later served as the first president of the Aircraft Owners and Pilots Association, wrote that his friend Cornish had a unique ability to smoothly handle the disputes that came before the board, which often involved witnesses of fairly high rank. "I always felt he had a sort of laid back Will Rogers touch in extracting agreements and ending arguments. He had a twist of humor and gentle persuasion and was almost never rattled. The closest to uneasiness I ever saw in Clarence was some nervousness the first two or three times we went to the White House. We both concluded that Truman was much easier to work with than Roosevelt and that Truman almost always came to more sensible conclusions than Roosevelt who was politically motivated beyond the real merits of the case being resolved." Interestingly, Cornish himself never mentioned those visits to the White House.

For Cornish, the World War II experience had been rich in hands-on opportunities to deal with serious problems in a working environment with men and women he respected and admired. Through his wartime service, he had gained self-confidence, increased his understanding of broad national issues, and honed his skills in working equitably and fairly with others at all levels of society. During those four years, he had flown to the far reaches of the nation. His horizons had expanded both literally and figuratively. He was now even better prepared for the next phase of his life: helping to shape Indiana's future in aviation.

10

Culmination of a Life in Flight

CLARENCE CORNISH HAD BEEN a pilot since 1918, and his involvement in Indiana aviation dated back to the mid-1920s. It had not taken him long to become an experienced participant and leader. Over a twenty-year period, he had gained the respect of the civil aviation community for his competence, foresight, and vision. Though the war had interrupted his career administering private and commercial aviation interests, his military service had been invaluable. His diverse activities and his connections in Washington had enabled him to gain an understanding of the inner workings of private, commercial, and military aviation, as well as federal and state legislative branches and agencies. He had been actively involved in all applications of the processes that affected aviation. It all added up to make him the logical choice to lead Indiana into the new air age.

That leadership would begin with the establishment of a state aeronautics commission, something that he had been trying to achieve for almost a decade. The constructive efforts that would finally lead to its formation began before the war. The Indiana Aircraft Trades Association requested that Governor M. Clifford Townsend appoint a committee to study and report on the needs of Indiana in the development of airports and other aspects of aviation.[1] On June 22, 1939, Townsend, who believed that aviation would become a vital economic factor in the lives of Indiana's citizens, appointed Cornish to head the first Governor's Fact-Finding Committee on Aviation. Serving with him were Lawrence Aretz of Lafayette, Marshall Kerr of Terre Haute, Clifford Potts of South

Bend, and I. J. (Nish) Dienhart, Adjutant General Elmer Straub, and Walker Winslow, all of Indianapolis. As commission chairman, Cornish traveled the state to gather citizen input and rally support for his vision of a vibrant aviation network for Indiana. "I wanted an airport in every county in this state, particularly at the county seat, if possible, and in the state parks," he said in 1992.[2] The committee's final recommendation, which was the culmination of a year of hearings held throughout the state, was to be placed before the Indiana legislature on January 9, 1941.[3] The twenty-two-page report focused on five major areas: (1) aviation development and promotion; (2) legislation, including the critical need for adequate zoning and protection of the approaches to airports; (3) an airport and airway plan; (4) private aviation; and (5) a state-level aeronautics department.[4] The board envisioned a department that would operate with a volunteer board and one paid administrator, to be funded through a tax on aviation gasoline at the rate of one cent per gallon.[5] By then, however, Governor Townsend had lost some of his initial enthusiasm, and he ended up shelving the report. Cornish was disappointed and puzzled. "I was so goddamn mad," he recalled. Townsend's dismissive reaction, of course, did not end the need to build support and infrastructure for aviation in Indiana. Although the 1941 initiative was not implemented, it laid the foundation for future studies and prepared the way for the inevitable growth of aviation in the state.

Even as World War II raged on, Townsend's successor, Henry F. Schricker, recognized that Indiana must prepare for the postwar era of flight. In 1944 he appointed Herschel A. Hollopeter to chair a new study group, the Governor's Commission on Aviation. During their deliberations, the group's twenty-four members almost certainly factored in the views of federal strategists, such as those expressed by David A. Postle of the Civil Aeronautics Board in an August 1944 memo, "Federal versus State Aviation Regulation."[6] He noted that there had been a flurry of legislative activity at the state level: in fiscal year 1943–44, 1,000 bills related to aviation had been introduced into forty-four state legislatures, of which 230 had subsequently been enacted. The result had been rampant confusion, with overlapping regulations, a lack of coordinated planning, and a disposition toward expanding aviation taxation. As Postle pointed out, "air traffic knows no state boundaries." He added, "All regulations governing air traffic must be of a national scope and must not be broken down into many diverse editions of state regulations."[7] Few would have disagreed that the rules and regulations covering the use of airspace and the certification of aircraft and airmen needed to be uniform throughout the country.

> **Cornish's Civic Service Record**
> **1920s–1940s**
>
> Fort Wayne Flyers' Club
> Fort Wayne Aviation Club
> Charter member of the Fort Wayne Aero Club
> Appointed delegate by Governor Paul V. McNutt to the National Association of State Aviation Officials (NASAO) in Detroit in 1935; appointed by McNutt as one of three permanent state delegates; chaired a Legislative Committee; served as president in 1947[8]
> Fort Wayne Chamber of Commerce Aviation Committee
> Indiana American Legion Aviation Committee
> Helped organize the Indiana Aircraft Trades Association; president, 1937
> Indiana Governor and member of Board of Governors, National Aeronautics Association, 1941
> Regional Director (eight states) of the American Road Builders Association (ARBA), Airport Division, 1940–41; presented a talk on "Airport Control" to the general session in New York City, 1941 (the stated objective of this group was the furtherance of legislation contemplated to provide federal aid for the development, construction, maintenance, and operation of a national system of airports)
> Executive Secretary and Treasurer of the American Association of Airport Executives, 1939–41 (letter of thanks dated April 1942 from AAAE, which was founded in 1928 and made it through the lean times in the 1930s "on the backs of a few dedicated volunteer leaders"[9])
> Chaired Governor Townsend's State Aviation Fact-Finding Committee, 1939–40 (tasked with answering questions about how Indiana might best equip itself with laws to cope with the needs of a growing aviation industry and develop a program to foster the growth of aviation; the committee's conclusions became the basis for the Aeronautics Commission Act in 1945)

Indiana was certainly no different from other states when it came to deliberating internal issues related to aviation, though several states had already established aeronautics agencies in one form or another. In a 1944 report, the Commission on Aviation concluded that the state needed to improve and expand its facilities for commercial air service and to promote the development of airports in smaller communities for the benefit of private pilots and businessmen.[10] Indiana had sixty-three airports at that time, but the Civil Aeronautics Administration's national plan called for twice that many within ten years.

Cornish kids around with Bill Slee at the American Road Builder's Association's annual meeting in New York City, 1939. (Author's collection)

Indiana's General Assembly took the CAA's plan and the new commission's report seriously. In March 1945, even before the war ended, legislators passed the Municipal Airport Act, Indiana Code 8-22-2, which granted municipalities the legal authorization to acquire, establish, construct, improve, equip, maintain, and operate airports and landing fields.[11] When the measure was signed into law, it represented the fruition of years of hard work by private flyers earlier and then by the members of the Commission on Aviation, who represented corporations, airlines, private pilots, municipal aviation boards and executives, cargo carriers, the state Chamber of Commerce, the National Guard, Purdue University, and the Civil Air Patrol, among others.

An arm of state government would now be needed to coordinate such development, and on March 28 the General Assembly passed the Aeronautics Commission Act. At long last, Indiana would have a state-supported agency that would deal specifically with aviation matters. The new Republican governor, Ralph F. Gates, appointed the members of the new Aeronautics Commission of Indiana, selecting George W. Starr, a professor of public utilities and transportation and director of the Bureau of Business Research at Indiana University, who had been an Army pilot during World War I; Morrison A. Rockhill, an attorney with wide experience in the field of aeronautics; and three private pilots: Gene Dawson, Charles L. Egenroad, and Guy T. Henry. The commission was

The members of the Governor's Commission on Aviation as of December 20, 1944, were as follows:

Herschel A. Hollopeter, Chairman
  Transportation Director
  Indiana State Chamber of
    Commerce
  Indianapolis
Robert H. McIntyre, Secretary
  J. D. Beeler, V-Pres. & Gen. Mgr.
  Mead Johnson Terminal
    Corporation
  Evansville
Frank W. Bodwell, Traffic Mgr.
  American Airlines Inc.
  Indianapolis
L. Hewitt Carpenter, Exec. Secy.
  Commission on Interstate
    Cooperation
  Marion
Dr. Beaumont S. Cornell
  Fort Wayne
John Dyer, Dist. Traffic Mgr.
  TWA Inc.
  Indianapolis
Charles L. Egenroad, President
  St. Jos. Co. Bd. Aviation Comm'rs
  South Bend
Theodore Ehrhardt, Chairman
  Kokomo Airport Committee
  Kokomo
Kenneth B. Elliott, V-Pres.
  Studebaker Corporation
  South Bend
Roger Fleming, Dir. Public Relations
  Allison Div., General Motors
    Corp.
  Indianapolis

Fred M. Gillies, Gen. Supt.
  Inland Steel Company
  East Chicago
Samuel C. Hadden, Chairman
  State Highway Commission of
    Indiana
  Indianapolis
Donald W. Hart, City Manager
  Eastern Air Lines
  Indianapolis
Frank McCarthy, Chairman
  Associated Railways of Indiana
  Indianapolis
Robert Prox, President
  Fran Prox Company
  Terre Haute
E. B. Reeder, Commercial Agent
  Railway Express Agency
  Indianapolis
Edward F. Rodefeld, Manager
  Rodefeld Company
  Richmond
Clyde S. Shockley, V.-Pres and Mgr.
  Muncie Aviation Corporation
  Muncie
Burton Swain, President
  National Veneer Company
  Seymour
Col. Roscoe Turner, President
  Roscoe Turner Aeronautical Corp.
  Indianapolis
Grove Webster
  Purdue University
  West Lafayette[12]

"empowered and directed to encourage, foster and assist in the development of aeronautics in this state and to encourage the reasonable establishment of airports, landing fields, and other navigation facilities."[13] The bill that created the Aeronautics Commission of Indiana defined several means to this end, requiring that the commission coordinate any plans for the state's airport system with the national plan, provide technical advice as needed, recommend pertinent legislation, approve or disapprove land purchases for airports, enforce all aviation-related rules and regulations, cooperate with the federal government, and implement the placement of air markings. The commission also had regulatory powers, enabling it to protect and ensure the interests of the public and promote safe aviation practices. In acknowledgment of the value of user input, two advisory councils would be created within a few months after official establishment of the Aeronautics Commission, one to be a clearinghouse for issues related to private flying including "co-ordinating programs for airport planning, legislation and safety" and the other to be concerned with all aspects of commercial aviation in the state, including "commercial airport development, legislation and flight and ground safety"[14] as well as airline operation. There were ten districts statewide, with one district at large.

Even before the legislation was adopted in March, 1945, the search for a full-time director had begun. Dr. Beaumont Cornell of Fort Wayne, who served on the search committee, sent feelers to Cornish about his interest and availability. Once the legislation had passed, Hollopeter, who was originally from Huntertown, Indiana, and who was the transportation director for the Indiana State Chamber of Commerce, would serve as acting director. In April, Cornell received a letter from Hollopeter suggesting that Cornish be given top consideration. "The Colonel impressed me as being one of the finest men I ever met," Hollopeter wrote. "I am very much impressed with him and from that I am entirely confident that he is the best man we could have gotten for Director of the Aeronautics Commission. Indiana owes you a great debt of gratitude for originating this idea [of tapping Cornish to serve as Director]."[15] Cornell responded, "Dear Holly: The more I think about it, the more I believe that Cornish is by all odds the best man in Indiana for this job. I have known him a long time. He is not a four-flusher in any way and has always backed up his convictions with action. Furthermore, he is very attentive to his work and he is 'hard' enough to see that those people who are working under him will toe the line." On March 13, Cornell sent a letter to Cornish to ascertain his interest. "If you're asked to take the job, I expect that you will turn it down, because I imagine that after the war you might be anxious to take over the airports here in Fort Wayne," he wrote, adding: "You are the logical man for this position, provided you want it[,] and I did not think we should look elsewhere until we had obtained your attitude."

The decision did not take long. On April 19, Cornish wrote Cornell to let him know that Gates had telephoned, and that "he wants me in."

> Governor Gates called me in Washington and wanted me to come out and take the job. I said, "Ralph (I knew him from up in Columbia City), I can't just tell General Arnold I'm going home. I have to get relieved from active duty to do that and I don't know at the moment whether that's what I want to do. I'll have to come out and see you. We'll talk about this. If this is just a law enforcement program, I don't want any part of it." He said, "Well come on out." So I flew out and met with Governor Gates and we had quite a nice session, but the state patronage man was sitting right there. I looked right over at the guy and said, "Ralph, there's one thing. If I even consider this further, I would have to have a distinct understanding that this is a non-political office, that I don't have to play politics to be the director of it. I've never played politics for a job in my life before and I don't intend to do it now." The Governor responded, "We'll take care of the politics, you take care of the business." So finally I decided to take the job, if I could get released.... There were other people that could have done it, but I appreciated it because I was the one who instigated it and I wanted to see it through. That was the job to promote the construction of adequate airports in this state, safety in aviation and the pursuit of aeronautics.

He added, "Apparently it's acceptable from Roscoe Turner's point of view." Turner, the well-known early barnstormer who had gone on to establish the Roscoe Turner Aeronautical Corporation in Indianapolis, presumably was also a member of the search committee.

In both Indiana and Washington, the wheels turned quickly and smoothly. On April 21, Cornish asked to be released from his duties in the military to "avail myself of the opportunity" to serve the development of aviation in Indiana. General Hap Arnold approved the request, effective July 5, 1945. In a June 19 press release, Gates announced his choice for director of the Aeronautics Commission: "The state is fortunate to secure the services of a man of Cornish's ability to head this new, important division of Indiana State Government." According to Richard L. Cunningham, who served as director of the commission from 1954 to 1962, Cornish was indeed the right man at the right time to focus on aeronautics in Indiana. The postwar years were an "explosive time," he said, with thousands of trained pilots having returned to civilian life, but there was not much opportunity in civil aviation. Before the war, there had been no impetus in this regard, but now there were large numbers of former wartime pilots who wanted to keep flying.[16]

Cornish was still in Washington, wrapping up his military duties and arranging the family's move back to Indiana, but he stayed informed about the Aeronautics Commission's early work. The first meeting was held at the statehouse on May 24, 1945, in the offices of Governor Gates. Present were Chairman Starr, commission members Henry, Egenroad, Rockhill, and Dawson, and Acting Director Hollopeter. Judge Harry Crumpacker of the Indiana Appellate Court administered the oath of office. The commission members left soon afterward to attend the initiation ceremony for the newly established Chicago & Southern Air Line at Weir Cook Municipal Airport.

Even before Cornish assumed the directorship, the commission began to taxi down the runway. At one of the first meetings, held in their offices in the Board of Trade building at the southeast corner of Meridian and Ohio Streets, the members began to consider issues that fell within their purview. The group's first official decision was made in support of an important link in the state's air chain, which provided reliable air transportation. The members voted in June to ask the Civil Aeronautics Board in Washington to grant a three-month extension to Eastern Air Lines for its route connecting Evansville to Chicago via Indianapolis. Chairman Starr sent a telegram to CAB chairman L. Welch Pogue with the request. A safety issue then arose in midsummer that required the commission's attention. When officials in Washington, Indiana, were told that they must remove or lower some wires and other potential flight obstructions before they could receive approval to build a new airport, they challenged the commission's authority. Hollopeter made it clear to Washington's mayor that they could not proceed with their plans until they were able to submit proof of a contract or agreement with the Rural Electric Administration to remove or lower the power lines on the south and east sides of the proposed site to a height approved by the CAA. Also required was written approval from the CAA stating that the changes met their flight safety standards. The Aeronautics Commission flexed its muscle. The message was firm and clear: safety trumped haste.

Cornish moved his family from Washington, D.C., to Indianapolis in August 1945. He was back in his home state, but not his "hometown, Fort Wayne," as he wrote wistfully to a friend in Texas, Ed Muir. However, he was ready to get to work. He took the reins of the commission that month. A Five-Point Program was soon adopted, designed to make flight accessible and safe for Indiana's citizens and to help them become contributors to future advancements in aviation. Cornish listed the individual points as part of a presentation for the CAB in the Great Lakes Area:

1. Development of the finest system of airports in the country.
2. Sponsorship, through local communities, of a complete air marking program covering the entire state.
3. Direction of an intensive air and ground safety program.
4. Fostering, through the state department of public instruction and local school officials, an aeronautical educational program extending from the grade schools through the colleges and universities.
5. Institution of a general program designed to develop and assist the growth of aviation, both private and commercial.[17]

Those five objectives served as a symbolic compass for Cornish throughout his tenure. As part of his job to familiarize Indiana's stakeholders with the benefits of the new program and the role that the Aeronautics Commission could play, he traveled to the far corners of the state and everywhere in between. On January 11, 1946, he was back in Fort Wayne as the guest speaker at an Aero Club meeting, where he had the opportunity to outline the Five-Point Program to some of the old friends and comrades whom he had flown with or worked with for the betterment of Fort Wayne aviation over a period of several decades.

It was not easy to implement the agreed-upon tasks with a small staff and an annual appropriation of $30,000, of which $6,000 was his salary. But Cornish ardently believed in the benefits of aviation, and he wanted every resident of Indiana to have access to the larger air transportation network, a point on which he never wavered. Ideally, each of Indiana's ninety-two counties would have its own airport, or at least one within fifteen minutes of its county seat. "If your combine breaks down in the middle of a harvest," said Dick Cunningham in 2009, "you want it up and running by nightfall. You call Chicago and have the part dropped off near the county seat, fifteen minutes from your farm. It meant a lot then and means as much now."[18] "You may call air land areas whatever you wish—airports, airparks, air harbors or flight stops, but they are community enterprises," Cornish quoted Pogue, who was now the president of the National Aeronautic Association, as having written.[19] "You cannot fly there if you cannot land there . . . and I believe that this should be posted on the walls of every City Council and every civic organization. Communities which disregard it or postpone it until 'sometime later' are inviting new business potentialities to fly overhead to the next town." It was a fine idea, but one impeded by stumbling blocks—public resistance, a "not-above-my-backyard" mentality, and concerns about practicality and whether the benefits would outweigh the costs.

Undoubtedly influenced by his pre–World War II exposure to the NAA's emphasis on educating youth, Cornish agreed with others that building an understanding of the value of aviation and its many aspects, both practical and pleasurable, would help to reduce public resistance. Believing that adults are often influenced by their children, he looked first to the schools, developing an air-age education program that was designed to build familiarity and promote enthusiasm among both the students and their teachers. If teachers had a better understanding of aviation subjects, he felt, they could apply this knowledge to their classroom work and help prepare their students to meet the technical needs of America's progress in aviation. He traveled throughout the state and talked with educators, who in turn submitted ideas to the state superintendent of public instruction, Dr. C. T. Malan. Together they developed a curriculum to integrate aeronautical materials into existing subjects. The resulting forty-seven-page guide, *An Air-Age Education Program for the Elementary and Secondary Schools in Indiana*, was distributed to the attendees at a teachers' training forum at Indiana University in the summer of 1947.

In 1948, Cornish brought George J. "Curly" Clingman on board to head the education program. Curly was an innovative jack-of-all-trades, a pilot and aircraft engine mechanic who had even tried wing walking. One of the first things he did was organize airport operation institutes for teachers in South Bend, Bloomington, and Evansville, "where instruction was provided by local flight operators in each area and a large percentage of the teachers were given actual flight experience."[20] When he spoke to groups or wrote about these institutes, he would point to the fact that when teachers were encouraged to visit local airports, it gave them the opportunity to see actual demonstrations of airway communications and control tower procedures, learn about the weather service, and understand the workings behind airline reservations and operation, airport management, fixed-base operation, and private aviation. The pillar of the program was a Link Trainer flight simulator, on loan from Link Aviation in Binghamton, New York; it was part of a mobile unit that also included visual aids such as charts, posters, and maps. The mobile unit, jointly sponsored by the Aeronautics Commission and the State Department of Public Instruction, helped educators come up with new ways to incorporate air-age materials into various courses, from mathematics, science, and geography to grammar and even home economics. There were so many requests for this popular exhibit that the staff could not meet them all. In the first half of 1949, 900 teachers and 15,250 students in Indiana schools attended demonstrations of the mobile unit. Visitors to the Indiana State Fair that summer also had a chance to see it and try out the Link Trainer. Unfortunately, despite widespread enthusiasm for the program on the part of both school authorities and teachers, the General Assembly did not authorize a continuation of funding for it, and the mobile unit was discontinued.[21]

Elizabeth "Betty" Pettitt (later Nicholas) was another member of Cornish's staff who helped him with his educational efforts—and became one of his closest friends in the process. Betty was a flyer; a member of the Ninety-Nines, the International Organization of Women Pilots, she had served in the Women Airforce Service Pilots (the WASP) during World War II, testing AT-6 training aircraft at Napier Field in Dothan, Alabama, where she stayed on after the war to work as a Link Trainer instructor. She moved to Indianapolis in 1947, where she flew as a skywriter in a specially equipped AT-6 before joining Cornish at the commission.[22] Her positions there were chief of special services, pilot, and editor of the departmental publication *Indiana Aero-Notes*. She was an enthusiastic promoter of aviation for the rest of her life. In 1952, 1981, and 1995 she was awarded the Indianapolis Aero Club's Dee Nicholas Trophy, given annually to the year's outstanding woman pilot.

Aero-Notes was published monthly from 1948 to 1965 and had a circulation of approximately four thousand. It was full of helpful information for pilots, covering topical subjects under headings such as "Air Marking," "Air-Age Education," "Airport News," "Accident Review," "Airport and Aircraft Fire Prevention," "Aviation Weather Services," "Civil Air Regulations," "Aircraft Maintenance," and "Activities of Flying Farmers and Local Flying Clubs."[23] In the October 1948 issue, a rather unusual notice alerted pilots that hunting deer from the air was prohibited. A more practical announcement let readers know about a 16 mm Technicolor sound movie, *Our Town Builds an Airport*, which could be obtained through the CAA District Airport Engineer's Office, 360 Massachusetts

Avenue, Indianapolis.[24] The film depicted the complete development of a municipal airport under the federal airport program and was available to groups that were looking to establish an airport in their own communities.

There were other ways to convince the public of the value of airports and airplanes. One was to demonstrate the potential for humanitarian service. On July 15, 1951, Cornish was one of hundreds of people who observed a simulated air attack by enemy bombers in Randolph County. The county's civil air defense unit, composed of seventeen private aircraft, shuttled medical personnel to the scene and removed evacuees and "injured." "It was the finest demonstration of potential use of light aircraft I've ever seen," Cornish said. "It proves conclusively that these planes and pilots, when properly organized such as in Randolph County, are capable of rendering valuable assistance in the event of a major catastrophe."[25] His view was shared by other high officials in military and civilian defense.

Most citizens would presumably understand and appreciate the value of having facilities in the area to support aircraft that could come to the rescue in emergencies such as that mock raid. But citizen landowners are understandably more apprehensive when they find out that a private or commercial airport is going to be constructed nearby, particularly when it is necessary for the developer to buy part of a landowner's property or to remove trees or destroy buildings and other structures on the property that might constitute a hazard to approaching and departing aircraft. In such cases, however, Indiana's Municipal Airport Act of 1945 does prescribe a remedy, ensuring that affected citizens will be compensated "for real property rights appropriated" and that they "have the same rights to procedure, notices, hearings, assessments, and payments of benefits and awards as are prescribed by statute for the appropriation and condemnation of real property."[26]

The question of who has the right to control the airspace above an individual's property led to a lawsuit that went all the way to the Supreme Court in 1946. A North Carolina chicken farmer named Thomas Lee Causby contended that the noise from a nearby airport being used by the military had literally frightened some of his chickens to death, putting him out of business. "By flying planes in this airspace, he argued, the government had confiscated his property without compensation, thus violating the Takings Clause of the Fifth Amendment." The U.S. Court of Claims agreed, and the government was ordered to compensate Causby. When the case went before the Supreme Court, the justices sided with the farmer. "If the landowner is to have full enjoyment of the land," read their ruling, "he must have exclusive control of the immediate reaches of the enveloping atmosphere." "Without defining a specific limit, the Court stated that flights over the land could be considered a violation of the Takings Clause if they led to 'a direct and immediate interference with the enjoyment and use of the land.' Given the damage caused by the particularly low, frequent flights over his farm, the Court determined that the government had violated Causby's rights, and he was entitled to compensation."[27]

Noise abatement was a real concern, one that Cornish addressed during a speech to the Indiana Aviation Conference in 1949.[28] He told the audience that while conducting tests the previous April on four prototypes of the new wind or castering landing gear, which allowed planes to land directly across a wind as high as forty miles per hour, the CAA had discovered that the use of these gears had an unexpected and positive side effect: by obviating an airport's need for multidirectional runways, they made it possible to reduce the noise level. With wind direction removed from the equation, aircraft outfitted with the wind gear could be safely accommodated by a single runway. And if that runway were constructed parallel to the nearest community, planes would not need to fly over congested or populated areas during takeoffs and landings. This added benefit aside, Cornish agreed that single-strip airports had an advantage over multidirectional airports, in that it cost less to construct and maintain their runways. His point was reinforced on November 25, 1949, when "CAA's Administrator enunciated the 'single runway policy' covering the use of Federal matching funds in the Federal-aid airport program. In substance, the new policy stated [that] additional runways that provided only wind coverage or conveniences

The *Indianapolis News* features Betty Pettitt Nicholas, shown with the Aeronautics Commission's Cessna 170, outfitted with a crosswind or castering landing gear. ("Air Secretary Works with Head in the Clouds: Woman CAP Colonel Has New Field to Conquer," *Indianapolis News*, 1952)

without increasing traffic capacity were not of sufficient value to justify the cost of construction."[29] Airports with single-strip runways could also welcome more regular airline service, an advantage for communities with limited budgets for airport development.

Hangared at Weir Cook Airport in Indianapolis in mid-1950, a Cessna 170, flown by Cornish and Betty Pettitt, replaced the commission's Stinson Voyager 150. The Cessna was equipped with the crosswind gear. Cornish once landed at Weir Cook with Edwin A. "Ed" Joyce, who was the CAA's supervising safety agent, and two members of the Aeronautics Commission as passengers.[30] When a strong wind switched from northwest to west, Cornish let the crosswind gear show its stuff. Joyce had never been in a plane with this gear, and he was not aware that it allowed the pilot to take off and land at a 35-degree angle before straightening out. When Cornish retold the story, he chuckled when recalling that Joyce panicked and yelled, "Straighten the plane! Straighten the plane!" as they touched down at an angle for a perfect landing.

Cornish also talked in his speech about another aircraft with what he called "good neighbor" characteristics: the Helioplane. It too offered the advantage of less aircraft noise. In 1948 Dr. Otto C. Koppen of the Massachusetts Institute of Technology and Dr. Lynn Bollinger of Harvard University had developed and built a prototype, which "was demonstrated by taking off from a tennis court at Harvard." The designers envisioned "an aircraft that could both land and takeoff on a short landing strip, provide safety features not found in other general aviation aircraft and be quiet enough to operate on airstrips adjacent to residential areas."[31] Cornish would undoubtedly have jumped at the chance to try one out, although there is no record that he ever got that opportunity. Neither the Helioplane nor the crosswind landing gear was broadly accepted for routine usage. In the end, airport neighbors had to cope and hope—cope with negative side effects and hope for the promised benefits of a local airfield.

Farmers generally did not need to be sold on the efficacy of airplanes or the need for runways. Mowed sod landing strips with windsocks popped up in fields everywhere. The agricultural use of aircraft usually evokes an image of a low-flying crop duster, but farmers found many other uses for their planes. As writer Jack Reese put it: "Those pictures of a plane dashing over a field at 15-foot altitude certainly give a distorted idea of how a Flying Farmer uses his plane. As to how the planes are really used—fence checking, obtaining emergency spare parts, getting data for contour plowing and spotting livestock ranked up toward the top, with crop-dusting a relatively minor use."[32]

In 1945, Cornish had stressed the need to establish more air parks and sod airfields to provide facilities for the thousands of personal aircraft that would soon be filling the skies.[33] By 1947 the Flying Farmers of Indiana, a group of private pilots, already had more than eight hundred members, and Cornish, an avid supporter of the organization, often dropped in on their fly-in breakfasts or fish fries. On April 24, 1948, he was one of some 550 people who attended a fish fry in Elwood sponsored by the Flying Farmers of Madison County, which was reportedly the largest "wing group" of Flying Farmers in Indiana, with 135 members.[34] During a speech at Purdue University later that year, Cornish recalled nostalgically that when he was operating Smith Field in the 1930s, he could easily name all the fixed-base operators in the state. There were 32 in early 1940s, a number that had grown to 207 by the end of 1948.[35] "We were like a small family," he said. "The pilots in the state could each be called by name, and at each outing almost all of them were sure to be there no matter where the event was to take place." For him, those were the good old days. Being with the Flying Farmers was the next best thing.

Cornish thrived on the fellowship of this group, and in the late 1940s he encouraged ways to kindle that kind of camaraderie among other flyers, re-creating the excitement and fun of the early days for the next generation of pilots. Under his tutelage, the department began to sponsor air tours. In 1947, a fall air cruise kicked off on Friday, October 17, at Turkey Run State Park, with a songfest and color slides of the park following dinner at the inn. The next morning, after an early breakfast, the pilots flew to Columbus Municipal Airport, "partaking of coffee and doughnuts furnished by the Columbus Chamber of Commerce" before traveling by bus to Brown County State Park. "Lunch was served in one of the Park's beautiful shelter houses after which the group returned to Columbus, said goodbye, and left for home."[36] The following spring, the department sponsored a tour in conjunction with the dedication of Wright-Patterson Air Force Base in Dayton, Ohio. Flying between Indianapolis and Dayton, a mass caravan retraced one of the first air-marked routes in the country. An all-Indiana Aero Round-Up, held during the first week of July in observance of the Second National Air Tour Week, drew approximately four hundred participants, who competed for prizes donated by various aeronautical agencies.

In Cornish's mind, private flying was the lifeblood of aviation. "It would be pale and lifeless without that wellspring of enthusiasm," he had said back in 1945. By 1948, there were approximately twenty thousand pilots in Indiana (many of whom had been trained during the war), and reliable aircraft were available for training purposes. In Cornish's words, "Small airplanes became much more stable and mechanical failure or fatigue became the exception rather than the rule." A person could learn to fly with greater confidence than when training "in an old crate." The cost had become more reasonable, too. "Older pilots might remember paying $300 for ten hours of training," he said, while in 1948 "a complete, deluxe course, including ground school, and earning a Private Pilot Certificate [cost] between $365 and $500."[37] Government educational benefits for veterans also helped to increase the pool of trained pilots, something that greatly pleased him. Another plus for private pilots was that the number of sod landing fields was growing significantly.

But more flyers meant more accidents. In 1946, for example, fifty-three persons were involved in aircraft accidents. Some were caused by the antics of what Cornish called "Johnny Buzz Boys"—pilots who insisted on flying at low elevations to impress their families and friends. *Indiana Aero-Notes* reported on one such incident: "The pilot was in the process of buzzing a house when a tree got in his way. Unfortunately, he was so intent in the buzz job that he didn't see the tree. He sustained serious injuries in the subsequent crash. Another incident of reckless flight."[38] Safe practices were a common topic in *Indiana Aero-Notes*. It included instructive articles about flying and landing in potentially hazardous conditions. One example was a warning to pilots to beware of landing on frozen snow because the plane's wheels could break through the crusty surface and cause the craft to flip onto its back.[39] Fortunately, airports that could afford field inspectors were able to quickly report such hazardous conditions to pilots, thus avoiding some serious outcomes.

Cornish and the department were quick to respond to concerns about the growing number of accidents. In cooperation with the CAA and CAB, they inaugurated the use of the Indiana State Police to investigate aircraft accidents and report violations of civil air regulations and state aeronautics laws. Additionally, the commission had a new air safety program in place by mid-1946, which had been developed with input gathered from airport officials at ten regional conferences.[40] Cornish then traveled around the state to inform airport managers about the program, urging them to implement its recommendations. Local personnel appointed an aviation safety director, who in turn formed an aviation safety committee. Having local safety committees that could promote safe practices both on the ground and in the air proved to be effective. Fifty-nine airports adopted the plan in 1947, and within two years, fatalities from civil aircraft accidents and convictions for infringements of Indiana's reckless flying laws had dropped significantly.

Air marking was still being used to guide pilots to airports, as it had been since the 1920s, but even though it seemingly should have been a simple matter to get the name and coordinates of each designated town painted on a single local rooftop, the process was still far from complete. In 1947, the CAA had allocated $1,600 to the state for paint, but only 48 of Indiana's 390 communities with a population above five hundred had complied with the directive. It would have been arduous for Cornish and his small staff to enforce that compliance on their own, but others volunteered to help with the task. Civil Air Patrol members acted as field representatives, gathering information. Boy Scout Troop #45 painted the marker at Deer Creek, after which the private pilots in the local Flying Farmers group rewarded the boys with free airplane rides. In towns such as Columbus, Connersville, Goshen, Martinsville, and Mt. Vernon, members of the Junior Chamber of Commerce (Jaycees) pitched in and implemented the needed air markings. Despite everyone's best efforts, the undertaking fell far short of the goal of 390. Cornish requested $10,000 from the 1949 General Assembly to fund two hundred air markings during the biennium. The request was approved, and the Airmarking Company of Rochester, Indiana, completed that phase of the task.[41] The department most likely never achieved full compliance, but it became less important as electronic navigation systems became more common. Dick Cunningham said, "As is often true with innovations considered crucial at one point in time, air markings lost their value by the mid-1950s with the use of sophisticated navigational aids."[42]

While speaking to a group of airport and fixed-base operators in February 1948, Cornish emphasized the need for a comprehensive navigation system. Both visual aids and radio aids were "often inadequate," he said, "and that's a problem for us to solve." Visual aids included runway markings, boundary lights, contact lights, and beacons; radio aids enabled pilots to determine their bearings relative to the source of a transmitted radio signal. Both commercial and military aircraft were increasingly relying on the latter.[43] For small fixed-base operators, however, or where radio aids were not available, runway and beacon lights would continue to be essential, especially for night flying. Cornish did not want to see the surplus beacons "mothballed" when they could be used to increase the safety level for personal pilots, so he asked the CAA to make them available, on a leased basis, to airport owners who could not afford to purchase them.[44] By 1950, twenty airports statewide had surplus beacons in place, and when Cornish left the department three years later, the commissioners gave him the credit in their farewell letter, stating that under his leadership, "certain beneficial programs, such as the beacon light plan, devised and implemented by Colonel Cornish have benefited the state of Indiana at no cost to its people and has been followed on a nation-wide basis."[45]

But obtaining surplus equipment for existing small airports was only one part of Cornish's job. He also advocated tenaciously for funding for the construction of new airports in smaller communities. The development of smaller facilities, he believed, would enable the growth of personal flying and the development of a feeder system to connect to commercial airline routes. In 1947 the Fort Wayne Chamber of Commerce conducted a survey to determine the economic value of airports to the community. The conclusion was that all of Fort Wayne's citizens benefited from the city's active aviation presence, with financial returns realized from rentals, landing fees, and concessions; the salaries paid to those employed in the field and the subsequent use of that money for local purchases; and capital investments.[46] The state's smaller communities stood to benefit similarly from the construction of a municipal airport. The CAA had a plan in place, and money had been authorized for the purpose, but despite the good relationships that Cornish enjoyed with higher-ups, he harbored some skepticism (probably stemming from Fort Wayne's experience in the early 1940s) about the federal government's ability to come through on its promises. He was outspoken about his concerns, commenting that many of the airports developed under the Works Relief Program either had never been completed or had subsequently "vanished into the woods, brush, and agricultural land from whence they came." Nevertheless, the federal funds were there to be used.[47] There was some question about whether those monies should go directly to the municipalities or be channeled through state departments of aeronautics. Ultimately they were channeled through state aviation authorities, with the federally aided public roads system serving as a model for what could be accomplished with intergovernmental cooperation.

The process began when a municipality submitted its plans for an airport to the commission. If the proposal was supported by a justifiable statement of economic need and proof of adequacy of the site and subsequently gained the commission's approval, funds could be made available from the Federal-Aid Airport Program. Department staff member and airport engineer Gilbert Schmitz could then assist local communities. With federal assistance, the number of municipal airports in Indiana grew from fourteen in 1944 to thirty-five in 1948. Eleven of them could handle scheduled air service.

In October 1947, the commission considered Delphi's application for a municipal airport. Community officials wanted to acquire between seventy and seventy-five acres of land located two miles west-northwest of the Carroll County community on State Highways 18 and 39. Both the Indiana Economic Council and the CAA's district airport engineer had endorsed the suitability of the proposed site, an area immediately west of the Wabash River commonly known as Pittsburgh Hill. The commission approved the project, and the plan went forward. During that same fiscal year, Cornish encouraged the commission's approval of municipal airports for Michigan City and Wabash. A total of $2,777,530 was spent on projects statewide during that period, which included land acquisition and the construction of new runways, control towers, administration buildings, drainage systems, and access roads, as well as taxiway lighting, paving, and utility improvements.[48] Cornish was also on hand when the Bendix Field–St. Joseph County Airport's new terminal building was dedicated on June 11, 1949. The community's commitment and the efforts of some public-spirited individuals had led to the construction of a modern facility with impressive accommodations, and Cornish had nothing but praise for the result. He told those assembled for the dedication, "Your new terminal building is the finest of its kind in the nation conceived as a wise public investment in the future of aviation so vital to the welfare of the citizens of St. Joseph County and the State."[49]

During Cornish's years in Washington, he had worked to get military air bases sited near populated areas so that they could be used by the local communities for civilian purposes following the war. In fiscal year 1946–47, eight Indiana airfields that had operated as wartime military air bases were acquired by the War Assets Administration and assigned to nearby municipalities: Bunker Hill Naval Air Station in Peru, Columbus AAF Auxiliary Field, Converse Naval Auxiliary Field, Baer Field in Fort Wayne, Galveston Naval Auxiliary Field, Kokomo Auxiliary Airfield, St. Anne Field in North Vernon, and Freeman Field in Seymour.[50] The availability of federal and municipal funds, coupled with the transition of airfields from military to civilian use, was moving Indiana toward a better, more equitable network of airports that was beginning to match Cornish's vision. He must have taken great satisfaction in this accomplishment.

While the conversion of those former bases was an important step in making flying safer for private pilots and available to more of them, it also offered the potential to bring nearby communities closer to being part of the much-discussed air route network. On November 5, 1943, while the United States was still engulfed in World War II—and well before the Aeronautics Commission of Indiana was created—the Civil Aeronautics Board had approved the initiation of an experiment involving local and feeder routes. The air routes that existed at the time served relatively large population centers, but smaller communities also needed to be considered and included. The CAB responded by inviting applications for an extended domestic air network. In 1944 there were 233 applicants nationwide for new service to a total of 3,097 cities. Those pending proposals would increase the existing route miles from 45,254 to 365,468. Indiana was part of the surge. Indiana's Aeronautics Commission supported the idea wholeheartedly. In 1945, Cornish appeared before the CAB's Great Lakes Area hearing on the applications. He submitted a list of 121 Indiana cities and towns that wanted to be included in a feeder and trunk air route system, noting which airlines would serve which communities.[51] There were 3,427,796 people living in the state at the time, less than a quarter of whom were provided with direct air service; if the submitted applications were approved, that proportion would increase to 55.2 percent.[52]

Achieving such progress in commercial aviation was not easy. All of the various entities with an interest in aviation—from the CAA and the CAB, to the commercial scheduled flight industry (airlines), to charter and cargo companies and other nonscheduled carriers, to the governing agencies of large and small communities, to the state itself—had a hand in the industry's growth, but the results were sometimes mixed. Good things were happening, yet there were stumbling blocks along the way. Need and reality often did not match. Carts went before horses. Cornish later outlined some of these problems, summarized as follows:

- The CAB usually made its final determination at the end of three years, which was later acknowledged to be too short a time period to adequately conduct the experiment.
- Communities were sometimes certificated (approved by the CAB) even though they did not have adequate airport facilities.
- Communities that did have adequate airport facilities were sometimes left without service despite having requested it.
- Air carriers certificated to serve a town sometimes discontinued that service because there was not enough business to generate reasonable dividends or because the local airport's navigational aids were adequate only for daytime flights.
- Poor scheduling resulted in a lack of acceptance by patrons.
- Some carriers insisted on complete wind direction components and avoided airports with a single unidirectional runway.

- Existing airfields sometimes lacked adequate runways to accommodate the increasing size and speed of aircraft.[53]

Cornish took great pride in representing the Aeronautics Commission and the National Association of State Aviation Officials in Washington, where he could have a say about decisions that would impact the entire nation. As an active member of NASAO, he chaired its Legislative Committee in 1946 and testified before the Congressional Air Policy Board regarding the proposed rules and regulations of the CAA under the 1946 Federal Airport Act.[54] His words at that hearing reflected his core beliefs:

> The 26 or more domestic air carriers operating in the neighborhood of 1,000 aircraft carrying hundreds of thousands of passengers are an important segment of civil aviation. So likewise are the thousands of small commercial operators and private aircraft owners operating 50,000 or more aircraft for the benefit of over 300,000 pilots. Here is the real heart of aviation in this country. It is the proving ground for the air men who become our transport and military pilots. It is the bringing of the advantages of air transportation to millions of our citizens located in hundreds of thousands of square miles not served by the present scheduled air carriers or the major air terminals.
>
> The provincial city dweller cannot believe because he has not seen the use of the personal aircraft made by the Flying Farmers, or the air

Cornish was an active member of NASAO and served as the association's president in 1947. He is shown here with some other officials in Minneapolis in 1950. (Author's collection)

tourist, or the businessman who is beginning to find it profitable to adapt his business to the medium of transportation by air. For these reasons, we [NASAO] support the principle of fair and equitable allocation of funds of *all* phases of aviation under the present Airport Act.[55]

NASAO elected Cornish to serve as its president for one year beginning October 28, 1947. (Cornish and Dick Cunningham are the only Hoosiers to have served in that capacity since the association's founding in 1931.) In 1948, NASAO officials met in Indianapolis and drafted a report to submit to the Congressional Air Policy Board in Washington. The organization then was committed to fighting for continuation of the GI flight training program, the elimination of which would have cut the industry's income in Indiana by some $5 million and possibly would have closed many airports.[56]

In the fall of 1948, Henry F. Schricker was elected to a second term as Indiana's governor.[57] Cornish, a devout Republican, was now faced with the possibility of being replaced as director of the Aeronautics Commission by a member of Schricker's Democratic Party. Weir Cook Airport's Phillip Roettger was a strong contender; his reputation as one of the country's most respected municipal airport managers made him an excellent candidate. Many, however, advocated that Cornish be retained in this important position. Letters from his supporters flew through the postal system. Jerome Beeler of the Board of Aviation Commissioners in Evansville, for example, wrote that there was a lot to be done during that developmental phase in aviation, and "Colonel Cornish" was still needed at the helm to ensure the "proper administration of Aviation affairs." Cornish was reappointed for a second four-year term. He later recalled his association with Schricker with fond memories and held him in the highest regard (but not so high as to cause Cornish to change his political affiliation). That potential personal crisis was averted, and Cornish's work continued.

He confirmed that aviation development nationwide had not gone totally as anticipated. In one of his strongest talks, he told the participants in the First National Seminar on Feeder Airlines held at the University of Oklahoma in 1950 that despite the hopes that had been raised after the passage of the Civil Aeronautics Act in 1938, "We do not, under the present system, have any assurance that air transportation will accomplish its objective or fulfill its purpose as envisioned by the Congress. . . . A plan for the development of an air route pattern, bringing into reality a well-balanced and well-integrated air transportation system in all of its major aspects does not exist." He asserted that when the act referred to the "domestic commerce of the United States," it was meant to apply to every person in every community in every state of the country, and that the inherent advantages of air transportation should therefore be encouraged and developed for all people. His frustration was clear.[58]

There were other ongoing frustrations related to the workload and to the commission's budget, as Cornish revealed in a letter that he wrote in June 1950 to his friend from the war days, Dave Postle: "What with an inadequate budget, trying to get things accomplished with the CAB, CAA, the municipalities, fixed-base operators, airport managers, personal pilots, police courts, other state officials and others, it appears that I am on a merry-go-round and thus going nowhere. At least I think so at times." Funds, or rather the lack thereof, were a constant concern, especially when state legislators were seeking ways to trim the budget. In 1951, State Representative Charles T. Miser (R-Garrett) sent chills through the Aeronautics Commission and three other state agencies when he suggested that they be eliminated. The editors of the Michigan City *News-Dispatch* viewed that as a "Penny-Wise, Pound-Foolish" move, and a step backward. "Indiana's Aeronautics Director Clarence F. Cornish wastes no money on press agents or ballyhoo. Consequently, few citizens realize that the commission works hard at many things." The editorial enumerated a number of the commission's accomplishments:

> It helps cities to build airports wisely and correctly. It maintains a lively safety program and cracks down hard on wild fliers—the buzz boys who endanger innocent lives as well as their own. It undertakes difficult, complex studies of the state's aviation problems—the key to intelligent future regulation. It has airmarked 300 cities and towns. It has kept the state's aviation laws up to date, helped to map all the state's airports, helped to plan future air routes that will serve both large and small communities. This service is rendered at a small cost. The commission's budget for 1951–52 is less than $50,000—a small sum as state agencies go. . . . It has chosen to regulate aviation by persuasion and education more than by red tape and bureaucratic bossing. It has given the taxpayers full measure for its few dollars.[59]

Saved by cooler heads, Cornish continued to advocate for people living in smaller communities, who had even greater transportation needs than the residents of large cities because they had to travel farther for services and supplies. A comprehensive local air route system would enable each community to be able to enjoy the best air transportation service that its facilities, economic conditions, and geographical location would permit. The dilemma of providing air service for residents of smaller communities was not limited to Indiana:

> After the war all long hauls between many cities were taken over by four-engined equipment and the DC-3 got the "leftovers," small cities and low-density traffic on short distances. Such routes are the toughest to make a go of, regardless of the type of aircraft employed, for two important reasons: (1) Every stop costs money. Thus operation with short hauls is more expensive per mile or seat-mile, than the long haul. The shorter the haul, the more marked the trend. (2) Load-factors are notoriously low in this type of traffic, producing a low revenue yield.... We are therefore still confronted with the problem of how best to transport passengers over hauls of 50–200 miles, with an average hop of 75 miles, in an economic way, and in competition with private automobiles, bus and railroad.[60]

Although it might not be possible to achieve the ideal—a network of feeder lines that connected to the major trunk lines with scheduled flights—perhaps some successes could be realized. If the CAB certification period were longer than three years, Cornish contended, more communities would be able to justify spending public funds on adequate airport facilities, with Congress providing the funding for appropriate navigational aids. As it was, the 1953 version of the Aeronautics Commission's annually updated Map of Certificated Airline Routes in Indiana shows eight airlines, widely dispersed, with service to Gary, South Bend, Fort Wayne, Lafayette, Kokomo, Terre Haute, Indianapolis, Richmond, Bloomington, and Evansville.[61] American, Chicago

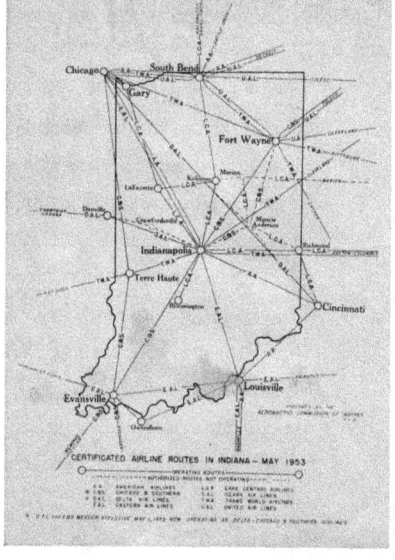

The Indiana Aeronautics Commission's 1953 "Map of Certificated Airline Routes" shows the state's scheduled airline service. (Author's collection)

When the first group of stewardesses completed their training for Lake Central Airlines in 1951, Cornish awarded them their pins. (Author's collection)

Cornish watches as the mayor of Indianapolis, Alex Clark, hands over the mail to be transported on the DC-3's final flight for Eastern Air Lines. (Author's collection)

& Southern, Delta, Eastern, United, Ozark, Trans World, and Lake Central were the active airlines at the time.

When Lake Central Airlines replaced its male attendants with female stewardesses in 1951, Cornish had the pleasant task of delivering a congratulatory speech—although not in an auditorium or another land-bound structure. The ceremony took place on June 30, while the fourteen young women and various officials circled above Indianapolis in a DC-3 Centraliner "Graduation Special."[62] "It is certainly a signal pleasure to have the opportunity of talking to you lovely young ladies who have just become official stewardesses of Lake Central Airlines," he told them. "This is certainly a historic day, having the honor and distinction of being members of the first graduating class of Lake Central Airline's stewardesses. By wearing the LCA wings, you become an intrinsic part of the airlines industry itself. . . . May God speed and watch over you throughout the air lanes that you travel in the future."[63]

Airline personnel have changed dramatically over the years since those stewardesses graduated in 1951, and so have airplanes. Cornish was on hand on January 31, 1953, when Eastern Air Lines made its last call in Indianapolis with DC-3 equipment; he and Postmaster George J. Ross watched Mayor Alex Clark hand the last sack of mail to pilot Captain H. T. "Dick" Merrill, President Eisenhower's campaign "air chauffeur."[64]

Change was not limited to the industry. What goes up must come down. That is true for airplanes such as the one that made the celebratory flight over downtown Indianapolis, and it is often true as well of careers. In the case of Clarence Cornish, that unfortunate reality led to one of his life's great disappointments. In 1952, Indiana elected a new governor, Republican George Craig. Again the position of director of the Aeronautics Commission became a matter of politics. Again supportive letters poured into the governor's office. Associates wrote that Cornish had earned nationwide recognition and was respected by members of the CAA and CAB, had done much for private pilots throughout the state, and was known for his integrity and civic interest. Even though he had been a lifelong Republican and despite the accolades, Craig appointed a West Point graduate as the Aeronautics Commission's second director, effective August 31, 1953. "The boys across the street lowered the boom on me. I am being replaced by Lieutenant David Krimendahl of Indianapolis—unknown and inexperienced in civil aviation," Cornish wrote to Roderick M. Wright.[65]

Knowing that his tenure was coming to an end, Cornish made his final rounds to check on a few of the myriad elements of civil aviation in Indiana that he had fostered. His 1953 desk diary for the first week in June reads:

> Monday: Checked air markers and visited airports at Connersville & Richmond;
> Tuesday: Checked air markers NE. Visited Baer Field and Tri State Apt. Hi Intensity lights inspected and OK. Tri State Apt. in very good condition as being developed by Paul Eyster. Business good at both airports;
> Wednesday: Checked air markers SW. Visited Airport at Vincennes and delivered Aerial Appl. forms to O'Neal Flying Service in connection with army worm infestation. . . . Sprayers doing everything possible to control pest but not enough planes & pilots;
> Sunday: Attended C. D. [Civil Defense] Fly-in Breakfast at Winchester this A.M. 750 served. Checked 12 air markers en route home.

After taking a helicopter ride with a Bell Aircraft Company pilot in Buffalo, New York, around 1946, Cornish returned to Indiana with a lasting enthusiasm for rotary-wing aircraft. (Author's collection)

With one week remaining in his term as director, the *Indianapolis Star Magazine* published an article on air travel in Indiana, with Cornish as the featured interview subject.[66] In this article, the last that would be written about Cornish's activities and views as director of the Aeronautics Commission for the State of Indiana, he talked about the growth in passenger travel, which he said currently represented about a third of the state's total flight activities. In 1952, scheduled airlines had carried 632,324 passengers who had started or completed their flights in Indiana cities, more than twice as many as in 1948, the first year that accurate records were kept. He also predicted a bright future for the use of helicopters in transporting passengers. It is not surprising that he spoke about this developing mode of transportation. Years earlier, around 1946 or 1947, Cornish had been the guest of Laurence D. Bell, who was born in Mentone, Indiana, and went on to be the founder and owner of the Bell Aircraft Company in Buffalo, New York. While there, he had the opportunity to ride in a Bell helicopter. Afterward he wrote his friend Dave Postle that he had returned to Indianapolis with a Bell Corporation–produced color sound movie titled *The Magic Carpet* and had used it on several occasions when speaking to civic groups and aero clubs throughout the state. "The helicopter develops a more keenly interested attitude and reception on the part of the audience than all of the others put together"—more interest even than supersonic speeds and supersized aircraft.[67] Right up until his departure from the commission, Cornish served as the chair of NASAO's Special Helicopter Committee and actively discussed "rotary-wing aircraft" (as they were still commonly called) with national leaders in aviation.[68] He believed that by the end of the decade, a single helicopter would carry thirty to fifty passengers and ferry them into the hubs with major airlines. He imagined that passengers, for the sake of convenience, would routinely be transported in short hops from flat-roofed buildings with heliports in Indianapolis, for instance, to Bloomington or Kokomo or other smaller communities, provided that they developed similar landing facilities.[69] However, his predictions regarding the practical applications of helicopter carriers for the convenience of passenger travel ultimately did not fly.

On September 1, 1953, Clarence Cornish joined the ranks of the unemployed. He received a letter, dated September 4, from the Aeronautics Commission of Indiana, chaired by George M. Beamer. The commission expressed gratitude for his "thirty-six productive years of pioneering in the aviation field." During his seven-year tenure as director, the number of public airports had more than doubled; airports and navigational facilities had made nearly $6.25 million worth of improvements; a beacon light plan had been devised and implemented at no cost to Indiana's citizens and had subsequently been replicated nationwide; and Indiana had become a leader in air safety. Others sent letters of praise and consolation, including Karl E. Voetler, an aviation development advisor for the CAA, who wrote:

> Clarence, you have done a grand job throughout your state and no one could have exceeded the efforts you put forth. Your results have been nothing short of continuous achievement and I well know they have been under great difficulties on many occasions. Those of us, like you and me, who have given practically our whole lives in the interest of the development of this business, deserve at least normal respect and consideration from those higher up. To have to operate entirely within political concept is nothing less than discouraging—to say the least. You will always have my highest regard and I hope you will not hesitate to permit me to express it whenever I can be of assistance to you.[70]

Such accolades may have brought Cornish some comfort, but they did not completely soften the blow. Weeks later, he responded to Voetler:

> Although my replacement was not entirely unexpected, it hit me pretty hard when I received the curt letter from the Governor's office. As a

matter of fact, I am still floundering a bit but hope to get things straightened out soon and get to work doing something to bolster the Cornish treasury.

In October he received a letter from the Flying Farmers of Prairie Farmer Land on behalf of the 2,500 private pilots represented by the club.[71] Signed by several friends, the letter said: "Aviation will always owe you a debt of gratitude. Whether it was a student pilot struggling for a private ticket, a Flying Farmer fighting encroachment, or a million dollar airline needing a voice, none ever found you wanting. The growth of aviation has been slow. It has needed cool heads and wise counsel, along with workers who would not shirk. On both a state and national level, you gave aviation all of this and more." In early December he responded in a letter to Bill Renshaw, "You must surely realize that I could not have done what I did without the continuing support of such grand persons and organizations such as the Flying Farmers. Their support and co-operation was the fuel that fed the fire which gave me the courage and inspiration to carry on toward our mutual objectives in spite of obstacles which at times appeared insurmountable."[72]

With the dust from a rough political scuffle settled and with the passage of time, Cornish could reflect on the opportunity he had had to work with others whom he respected, both locally and nationally, and to focus on his achievements. In his later years he often jotted down highlights from his life, mainly his professional career. He wrote the following about his time with the Aeronautics Commission[73]:

> Under my leadership as Director of Aeronautics, a statewide program for development of aviation in the State was developed and adopted by the Aeronautics Commission. This program involved, but was not limited to, the development of an airport construction and improvement plan; a flying safety program; an air age education program in the secondary schools; a pilot training program under the Veterans Education benefits of the Federal government, and [a] publication [and] appearances over the state.
>
> These activities involved sponsorship of amendments to existing laws and enactment of new enabling acts at both the State and Federal levels. ... This phase of our activities resulted in elimination of the old "Absolute Liability Act" regarding the liability so imposed on the owner and/or

operator of an airplane in Indiana, and enactment of the Federal Airport Act to provide federal funds for the development of public use airports in the United States. This was accomplished with the cooperative assistance of many progressive aviation minded individuals throughout the State and local, state aviation organizations such as local aero clubs, Indiana Aviation Trades Association, the Flying Farmers of Prairie Farmerland. At the national level, the efforts of the National Association of State Aviation Officials, in which I served as Legislative Committee Chairman and subsequently as President (1947), were most effective.

The Indiana Aeronautics Commission participated in many hearings by the Civil Aeronautics Board involving scheduled air carrier route and service matters affecting Indiana communities. When the Civil Aeronautics Administration announced the plan to discontinue operation of their airway beacon lights, we were able to convince them not to sell them for junk but to release them through the State aviation authorities, without cost, for acquisition by public and private airport owners to be used as airport beacons. Many of these beacon lights were assigned to and erected at Indiana airports. To aid cross-country flying, a statewide "air marking" program was vigorously pursued.

The president of NASAO, Joseph K. McLaughlin, wrote to commend Cornish for his "sound and aggressive thinking and work in the safety, scheduled air carrier, legislative and many other aspects [of aviation] . . . which have with credit borne the imprint of your handiwork."[74] This sentiment was not lost on Cornish, who responded, "I have given the better part of the past thirty-six years to aviation, both civil and military, and I sincerely hope that I may find some niche in which I can continue to serve in one capacity or another."

He was thankful for what this field he loved had given him, and he hoped to give back even more.

11

Never Call It Quits

Clarence Cornish was fifty-four years old. He had a level of expertise in civil aviation that few could match. But with nearly forty years of accumulated knowledge of the field, he was now out of work. With nothing on the horizon for him in aviation, he turned in a new direction. When friends Harold and Bill Wells, of the Wells Insurance Agency, invited him to join the firm to handle the aviation sector of their business, he decided to give it a try. There were not a lot of aircraft policies to be sold, but he found that he was good at the job, so in 1955 he bought his own agency, which he built into a "fairly nice business." He sold general casualty insurance for the rest of his working life. "When I reached sixty-five I was making too much money to collect Social Security so I just went ahead and worked." At the age of seventy-two he said, "I've had it," retiring for good in 1970.[1]

But he was not yet ready to retire from flying. After all, he had been piloting aircraft either through the military, private industry, or the state government since 1918. He was so passionate about it during those early years that he would happily have flown every day: in the summer of 1929, his Pilot's Book shows that he in fact did fly every single day for sixty-two days straight. He had never owned his own plane, however, so he now had to rent a Cessna 172 when he wanted to take to the air. But it was hard to justify the expense. "I can't afford to fly too often," he said at one point. "So I fly just enough to maintain a reasonable level of proficiency."

Fifty years from the date of his first flight with Billy Brock on May 6, 1918, he and a group of friends, family, and business associates drove to the Indianapolis Metropolitan Airport to mark the occasion. At the age of seventy, Cornish took off and landed a few times that day, taking two or three of his followers each time he went up for a brief loop above the northeast suburbs of Indianapolis. It was to become an annual salute to the event that had changed his life. In the coming years, he would underline and star each anniversary flight in his Pilot's Book. To continue flying, he needed to stay physically and mentally fit. His unswerving dedication eventually led him into the history books in a way that no one could have predicted—but that was still far on the horizon.

He was eighty for the sixtieth-anniversary flight, and afterward a luncheon was held in his honor. He had hoped that Helene Foellinger, who was still the publisher of the Fort Wayne *News-Sentinel*, would be able to attend the festivities, but she had to send her regrets. "Dear 'Cap,'" she wrote, "I'll be with you in spirit as I recall all the fun times we used to have in the good ole 'Yankee Clipper' and the times when you and Lois used to 'baby-sit' Loret and me. Have many more Happy Landings."

As the years rolled along, the number of media representatives who were on hand to chronicle the annual celebration grew. Longtime *Indianapolis Star* reporter Rex Redifer had first covered the story in 1974. On May 6, 1992, on the seventy-fourth anniversary, Redifer joined the crowd to watch "Cap" "roar off down the runway and circle back for a couple of low-level waggle-winged flybys."

Clarence cuts a commemorative cake following his sixty-fifth-anniversary flight in 1983. Lois and Ruth Ann are ready to assist. (Author's collection)

At the age of ninety-three, Cornish was now the second-oldest U.S. male pilot, having been born November 10, 1898. Harry Liggett of Limon, Colorado, was the oldest, born June 3, 1896.

Noah Adams interviewed Cornish for that day's broadcast of National Public Radio's popular news program *All Things Considered*. When Cornish was asked how he would know when the time had come to ground himself, he laughed. "How will you know? I'm going for seventy-five, which will be next year," he said. "And the good Lord willing, I'll make it." Adams wished him a good year and concluded, "He's reported to be the country's second oldest pilot, but Mr. Cornish tells us the other pilot has only been flying for thirty years."

Cornish was not a foolish man, and he did not take chances with his own life or the lives of others in his later decades. By the time he reached his ninetieth birthday, he always took along an instrument-rated copilot as his "insurance policy." "No use being silly about it," he said. "At my age, things can happen. My ticker goes out, the fellow in the other seat takes over." Most often it was Tim Crean who accompanied Cornish on those later anniversary flights.

May 6, 1993, marked a major milestone. On that day, Cap celebrated seventy-five years in flight. Once again, Betty Nicholas and Ruth Ann had planned a memorable luncheon event for the occasion, and Betty, in her efficient manner, had contacted the media, notified dignitaries, assembled mailing addresses, and sent invitations. Friends, relatives, and representatives of government gathered in the Indianapolis Metropolitan Airport's hangar. The accolades were abundant, and Cornish, attired in his traditional, understated knit collared shirt, bolo tie, cap and sneakers, was presented with his third Sagamore of the Wabash Award.[2] Three weeks later, on the morning of May 27, 1993, Cornish and Crean broke from their usual routine of taking a few spins over northeastern suburban Indianapolis. They headed south to Tennessee, to the site of the old Park Field, where Cornish had made his first solo flight in a Flying Jenny seventy-five years earlier. Captain Earl W. Shaut, the commanding officer of the Naval Air Station, and Captain Lanny Miller, the commander of air operations, were among a group of Navy personnel and news media representatives on hand when Cap taxied the Cessna 172 to a stop at the Millington base. During lunch he opened his Pilot's Book to show where he had registered his first time aloft in 1918 at that very location.

There would be two more May flights—in 1994 and 1995. After the latter, Cornish believed in his heart that his time in the air was over. He was as mentally sharp as a person decades younger, but his body was giving out. Cap was

Cornish's good friend Edmund F. Ball of Muncie helped him celebrate his seventy-fifth year of flying in 1993. (Author's collection)

Captain Earl W. Shaut, commanding officer of the Naval Air Base in Millington, Tennessee, welcomed Cornish back to the site of the former Park Field on May 27, 1993. (Author's collection)

winding down, but his granddaughter Christy Krieg had a different idea about how he should wind up his flying career. Having seen the record for the world's oldest actively flying pilot in the *Guinness Book of World Records*, she realized that her grandfather might be able to beat it. According to the Guinness documents, Cornish would break the record if he piloted a plane after July 21, 1995.

However, Cornish had already made his decision to quit flying, and initially he was not interested in going up that one last time. But in August he made a

In Cornish's Pilot's Book, he logged his first solo flight from Park Field on May 27, 1918. (Author's collection)

surprise announcement. "Well, I guess if we're going to do this Guinness thing," he said, "we'd better get started." Everyone sprang into action. Cornish reserved the plane, and Crean agreed to be the copilot. Betty Nicholas notified the media. On August 16, 1995, at the age of 96 years, 279 days, Cornish once again placed his hand on the yoke, taxied, and lifted off—breaking the world record by twenty-one days.

But there would be yet one more flight a few months later. A U.S. representative for a new show being launched in South Korea called Cornish to ask him whether he would be willing to pilot an airplane with the show's hostess, Miss Chou, and a cameraman on board. He declined apologetically, but the show's intermediaries persevered. On December 4, Cap briefly forgot his infirmities. Casting aside his cane, he escorted a Korean starlet across the tarmac to the waiting plane. He stepped up and into the pilot's seat, leaned down to kiss Lois goodbye with the news cameras rolling, and latched the door. With Crean to his right and the two Korean passengers behind him, he taxied out for his final takeoff. When he piloted the Cessna that day, he was 97 years, 24 days old. He had broken his own record.

* * *

Eighteen days later, Cornish "went West," having lived through all but five years of the Century of Flight. His long-time colleague and fellow pilot Betty Nicholas

Betty Nicholas, shown here with Cornish, was the primary planner behind the major anniversary events. (Author's collection)

South Korean TV hostess Miss Chou walks with Cornish on the day of his final flight, December 4, 1995. (Author's collection)

was the last person, other than immediate family members, to visit him during his five-day stay in hospice. Leaning over him, she said, "It's Betty. Want to go for a flight?" His final verbal communication: "You bet."

Betty had been his staunchest supporter over the years, and she wanted to make sure that he received one more honor. Through her efforts, he was named an Honoree in the International Forest of Friendship, a living memorial to aviation that overlooks Lake Warnock in Atchison, Kansas. In June 1997 he posthumously took his place among the other new inductees, including Indiana's Roscoe Turner and Mid Cassidy, astronaut Walter Shirra Jr., and astrogeologists Carolyn and Eugene Shoemaker, for whom Comet Shoemaker-Levy was named. Cornish's stone rests with those of other Hoosiers in the shade of a tulip poplar, Indiana's state tree.[3]

While Clarence Cornish began flying too late to qualify as an Early Bird of Aviation, he was nevertheless an aviation pioneer. Having first taken to the skies as one of the hundreds of "Jenny-come-latelies" who learned to fly during World War I, he had gone on to a career that took him from Fort Wayne, Indiana, where he was involved in every aspect of the development of private and commercial aviation; to Washington, D.C., where he made crucial contributions to America's defense during World War II as chief of the Flight Operations Division of the Army Air Forces; and then back to Indiana, where he worked with several governors to advance the development of aviation in the state. With a true dedication to the cause and a steady hand on the controls, Cap Cornish helped to lead the Hoosier state during the early evolution of the air age.

Cornish waves before taking off at the Indianapolis Metropolitan Airport on December 4, 1995. (Author's collection)

Cap's Last Flight

"Here he comes!" the angel said
As he swung wide the gate.
St. Peter smiled as Cap walked in
And said, "Sorry you had to wait."

"I had so many things to do
There wasn't time to leave;
And I was honored at the end,
Truly hard to believe."

We know the story very well,
The airports he helped create;
He ran a few, the Commission too,
And they're all doing great.

Every May he "took 'er up"
For seventy-seven years,
He proved that age had nothing to do
With outperforming his peers.

He flew in World Wars I and II
And served his country well.
He made many friends along the way
Before he heard the bell.

"You've more than earned your wings, my friend,"
St. Peter said with a smile;
"And don't give a thought to all those friends,
You'll see them after a while."

And so, my friends, don't mourn too long
Just look up to the sky,
He just may be that red-tailed hawk
Circling way up high.

—Betty Nicholas, 1996

RECOGNITIONS EARNED BY "CAP" CORNISH

"Cap" Cornish earned accolades during his seventy-seven active years in civilian and military aviation:

Who's Who in Transportation and Communication in 1942

Father of Fort Wayne Aviation by OX5 Aviation Pioneers, Indiana Wing, May 24, 1975

Commendation by Art Smith Aeroplane Society in 1978

Inductee, **OX5 Aviation Pioneers Hall of Fame,** San Diego, September 1986

Recognition by City of Indianapolis when Mayor Stephen Goldsmith proclaimed **June 9, 1992, "Colonel Clarence F. Cornish Day"**

Recognition as a **pioneer in the development of aviation as a means of transportation** by the Indianapolis Aero Club at a banquet in his honor held the same day

Three times **Sagamore of the Wabash**—1978, 1988, and 1992

Honored Founder Member at the 1993 banquet of the Order of Daedalians,* Air Force Museum, Wright Patterson AFB, Dayton

Indiana Aviation Person of the Year by the Aviation Association of Indiana in 1994

World's Oldest Actively Flying Pilot, *Guinness World Records,* 1997–2003

---

*The Order of Daedalians honors, as its founder members, all WW I aviators who were commissioned as officers and rated as military pilots no later than the Armistice on 11 November 1918. It perpetuates their names as the first to fly our country's airplanes in time of war.

Notes

CHAPTER 1

1. Immigration papers in author's collection.

2. Advertisement in *Kramer's General Business Directory: Containing an Accurately Selected and Classified List of the Leading Manufacturers, Jobbers, Wholesale and Retail Dealers, Professional and Business Men of Northern Indiana* (Lafayette, Ind., 1884), 345.

3. This letter and the announcement cited in the following paragraph are from the author's collection.

4. L. W. Wilson and L. R. Pfaff, *Early St. Marys: A History in Old Photographs from Its Founding to 1914* (St. Marys, Ont.: Stonetown Books, 1995).

5. Personal communication, 1995.

6. Author's collection.

7. Stephen Wallace Merrihew, "Automobile Possibilities," *Frank Leslie's Popular Monthly* 57 (1904): 353–74, quotation from 354.

8. Jay Spenser, *The Airplane: How Ideas Gave Us Wings* (New York: HarperCollins, 2008), 24.

9. Handwritten memoirs in author's collection.

10. Fred D. Cavinder, "The Daring Young Miss in Her Flying Machine," *Indianapolis Star*, Magazine Section, April 24, 1966, 14.

11. *Art Smith's Story: The Autobiography of the Boy Aviator Which Appeared as a Serial in the News-Sentinel*, ed. Rose Wilder Lane (1926; reprint, Fort Wayne, Ind.: s.n., 1975), 30.

12. William Randolph Hearst is considered the founder of the American Boy Scouts. Patterned after the British Boy Scouts, the organization emphasized the obligations and duties of citizenship.

13. The quotations in this paragraph are from Clarence F. Cornish Interview, Douglas E. Clanin World War II Oral History Collection, 1944–2002, Collection M 0783,

Manuscript and Visual Collections Department, William Henry Smith Memorial Library, Indiana Historical Society, Indianapolis [hereafter IHS interview]. Clanin was an editor with the Indiana Historical Society. Between 1983 and 2002, he conducted interviews with World War II veterans. The Clanin collection includes both the original audiotapes of the interviews and later transcriptions; the transcripts of Cornish's interview, which was conducted on January 11, 1992, are in box 4, folders 11–12.

14. Quotations from IHS interview.

15. IHS interview.

16. Boy Scouts of America: Anthony Wayne Area Council. Cornish's correspondence collection reveals that the members of his Scout troop stayed in touch over the years and got together annually, even into the 1940s. They called themselves "the Edwin T. Jackson Association of Old Scouts." Jackson was the "Daddy of Scouting" in Fort Wayne and organized Troop 1; J. R. Bird was the first Scoutmaster. Cornish wrote Carl Bullerman in April 1942: "For the second time, a World War separates us from our old ties and friends back home and I deeply regret that I will be unable to attend the 1942 Annual Meeting of the Old Scouts. . . . May we all hope and pray that this World Catastrophe is soon ended and that we may again resume our normal way of life with our old friends and among old and familiar surrounds."

17. Quotations from IHS interview.

18. Ibid.

19. For more information, see "Elementary Radio Publications," *Radio Service Bulletin* 60 (April 1, 1922): 16, with reference to *Elementary Principles of Radio Telegraphy and Telephony* (Washington, D.C.: GPO, 1921), which explains the fundamental principles in simple terms.

20. "Will Government Buy Fort Wayne Wireless Telegraph Stations," *Journal-Gazette*, February 13, 1916.

21. Personal communication, 1975.

22. "Fort Wayne Joins in Progressive March of Wireless Age with Its Rapidly Growing Radio Club and Many Stations," Fort Wayne *Journal-Gazette*, March 27, 1921, 43.

23. "Wireless Club Is Formed for Practice Here," unidentified newspaper clipping dated July 2, 1915; "Radio Club Formed in Fort Wayne by Wireless Enthusiasts," Fort Wayne *News*, July 19, 1915, 6; "Wireless Club Organized: Prof. G. W. Carter Is Named President of New Association," Fort Wayne *Weekly Sentinel*, July 21, 1915, 26.

24. "Will Government Buy Fort Wayne Wireless Telegraph Stations," *Journal-Gazette*, February 13, 1916.

25. "Wireless Fans Organize and Feast," *Journal-Gazette*, January 16, 1916, 28; "Radio Club in Its First Annual Banquet," unidentified newspaper clipping.

26. Cornish subscribed to *QST*. The monthly magazine of the American Radio Relay League, it cost ten cents per issue. An incomplete, typed letter dated March 16, 1917, from Cornish to the QST Publishing Company in Hartford, Connecticut, states

that he was enclosing a money order for a dollar for a one-year subscription to "the Best Wireless Periodical in the world," and that he always looked forward to receiving it. At some point he circled his call number, 9KG, in three places in the February 1917 edition where it was noted that he had been copied at Rochester, Pennsylvania, and heard by others.

27. "QRM" was the official code name for manmade interference, asking and responding to the question, "Are you being interfered with?"

28. *QST*, February 1917, 36–37, 47.

29. "Electro-Technics to Study Wireless: Ft. Wayne Radio Club to Give Demonstration of Marconi System," *Journal-Gazette*, April 5, 1917, 2. "The Electro-Technic club, of the Fort Wayne works of the General Electric company," was described in a full-page article as "an organization which handles the educational, social and athletic phases of the big General Electric works, located on Broadway." "Fort Wayne's Greatest Industry, the General Electric Works, Has Splendid Organization in the Electro-Techno Club: Its Social and Educational Features," *Journal-Gazette*, February 18, 1917, sec. 4, 1.

30. An article titled "Wireless Dance Held in Morristown N.J.," *QST*, February 1917, 47, tells about a regenerative system and a two-step amplifier attached to a loudspeaking phone.

31. *Compton's Pictured Encyclopedia*, Vol. 9, VIN-ZWI (Chicago: F. E. Compton & Company, 1929), World War: The Greatest War in History.

32. B. J. Griswold, *The Pictorial History of Fort Wayne: A Review of Two Centuries of Occupation of the Region about the Head of the Maumee River* (Chicago: Robert O. Law, 1917), 572, 574. The *Journal-Gazette* article is quoted on 572.

33. Ibid., 584, 585.

34. B. J. Griswold, *The Pictorial History of Fort Wayne: A Review of Two Centuries of Occupation of the Region about the Head of the Maumee River* (Chicago: Robert O. Law, 1917), 572, 574. The *Journal-Gazette* article is quoted on 582.

35. *Journal-Gazette*, May 14, 1917, 3.

CHAPTER 2

1. He kept the letters for the rest of his life; they were found in an unlabeled wooden box after his death. Author's collection.

2. Letter dated October 10, 1917.

3. Letter dated November 14, 1917.

4. Letter dated November 21, 1917.

5. Robert Wohl, *A Passion for Wings: Aviation and the Western Imagination, 1908–1918* (New Haven, Conn.: Yale University Press, 1994), 69.

6. Ron Dick, *American Eagles: A History of the United States Air Force* (Charlottesville, Va.: Howell Press, 1997), 17, 18, 20. A pusher refers to an airplane with its propeller

facing backward, mounted to the back side of the engine behind the cockpit. This arrangement pushes the airplane. A tractor-style design has the propeller facing forward and mounted in front of the engine; it pulls the airplane.

7. *Compton's Pictured Encyclopedia* (Chicago: F. E. Compton, 1929), World War: The Greatest War in History.

8. Ibid., 59–60.

9. Ibid., 60.

10. Dick, *American Eagles*, 21.

11. Rea N. Redifer, *Once upon a Canvas Sky* (York, Pa.: York Graphic Services, 1984), 49.

12. Ibid., 42.

13. Arthur Gordon, *The American Heritage History of Flight* (New York: American Heritage Publishing Co., 1962), 118.

14. "In the space of a year . . . armies of lumberjacks selected prime spruce, farmers planted acres of castor beans for lubricating oil; industrialists not only created factories for fittings, linen, dope, engines, instruments, wire, armament and all of the thousands of other items necessary for a combat airplane, but also trained the tens of thousands of workers to make them. Aircraft manufacturers seized upon existing designs in 1917, and were building at the rate of 12,000 or more aircraft per year by the end of the war in November 1918." Walter J. Boyce, *The Smithsonian Book of Flight* (Washington, D.C.: Smithsonian Books, 1987), 106.

15. Redifer, *Once upon a Canvas Sky*, 77.

16. Willis J. Abbot, *The Nations at War* (New York: Doubleday, Page & Co., 1918), 277.

17. Rebecca Hancock Cameron, *Training to Fly: Military Flight Training 1907–1945* (Air Force History and Museums Program, 1999), p. 108.

18. Letter from Cornish dated February 10, 1918.

19. Clarence F. Cornish Interview, Douglas E. Clanin World War II Oral History Collection, 1944–2002, Collection M 0783, Manuscript and Visual Collections Department, William Henry Smith Memorial Library, Indiana Historical Society, Indianapolis [hereafter IHS interview]. Clanin was an editor with the Indiana Historical Society. Between 1983 and 2002, he conducted interviews with World War II veterans. The Clanin collection includes both the original audiotapes of the interviews and later transcriptions; the transcripts of Cornish's interview, which was conducted on January 11, 1992, are in box 4, folders 11–12.

20. Quotations from IHS interview.

21. Tom LeCompte, "The Few, the Brave, the Lucky," *Air and Space Magazine*, July 2008, 48.

22. Cornish quotations from IHS interview.

23. Harry Golding, *The Wonder Book of Aircraft for Boys and Girls* (London: Ward, Lock, 1919), 86.

24. Entered as JN4"D" by Cornish in his Pilot's Book.

25. A. Scott Berg, *Lindbergh* (New York: G. P. Putnam's Sons, 1998), 70.

26. Cadet C. F. Cornish letter to Dear Mother dated May 27, 1918, from Park Field, Memphis, Tennessee. Author's collection.

27. "Pilot Takes to Air, Recalls First Flight," *Anderson Herald Bulletin* (Anderson, Ind.), May 8, 1988, Indiana Historical Society, Folder: Air Pilots—Indiana

28. Stephen Longstreet, *The Canvas Falcons: The Men and Planes of World War I* (New York: Barnes & Noble Books, 1995), 243.

29. Gordon, *American Heritage History of Flight*, 163.

30. Letter dated June 18, 1918.

31. Undated informal interview conducted by an unidentified person who was apparently writing a history of Waco aircraft. Transcribed from audiotape in author's collection.

32. Cornish's handwritten reports to Officer in Charge of Cross Country Flying. Author's collection.

33. IHS interview.

34. Original letter in author's collection.

35. Letter dated September 26, 1918.

36. Redifer, *Once upon a Canvas Sky*, 83.

37. The sources for the information in this paragraph are a telegram from Cornish dated October 4, 1918, and letters from Cornish to his mother dated October 4, 6, and 10, 1918.

38. Ibid.

39. Letter dated October 19, 1918.

40. IHS interview. The quotes in the subsequent paragraphs are from this interview.

41. Gene M. Burnett, *Florida's Past: People and Events That Shaped the State*, vol. 3 (Sarasota, Fla.: Pineapple Press, 1991), 7.

42. Letter dated December 16, 1918.

43. Wohl, *Passion for Wings*, 255.

CHAPTER 3

1. Welman A. Shrader, *Fifty Years of Flight: A Chronicle of the Aviation Industry in America, 1903–1953* (Cleveland, Ohio: Eaton Manufacturing Co., 1953), 25, 26, 30.

2. "Greater Development of American Aeronautics Urged at Atlantic City Convention," *Aerial Age Weekly*, June 9, 1919, 623; Shrader, *Fifty Years of Flight*, 25, 30.

3. Peter Garrison, "By Stars, Beacons, and Satellites: How We Got from Point A to Point B," *Air and Space*, March 2006, 50–57.

4. Lloyd B. Walton, "Cap Cornish—On a 'High' That's Lasted since 1918," *Indianapolis Star Magazine*, October 7, 1973, 10.

5. An article in the November 28, 1981, *News-Sentinel* begins: "The iron horse used to move the nation. And Fort Wayne's Bass Foundry moved the iron horse."
6. Unidentified newspaper clipping.
7. Ibid.
8. "Miami Athletic Club Banquets at Baltes," *Journal-Gazette*, June 8, 1919, 21. The Baltes Hotel was at 1123 Oakland Street.
9. "Two Local Boys Start for Canada in Canoe," *Journal-Gazette*, August 28, 1920, 2; "Start by Canoe to Canada," Fort Wayne *News and Sentinel*, August 30, 1920, 2.
10. According to the website of Kekionga Middle School in Fort Wayne, Kekionga was "the capital of the Miami Tribe. The village was located at the confluence of the Saint Joseph, Saint Marys and Maumee rivers, part of Fort Wayne today. Kekionga was an important trading post and was once called by Chief Little Turtle 'that glorious gate . . . through which all the good words of our chiefs had to pass from the north to the south and from the east to the west.'" http://kekionga.fwcs.k12.in.us/about.php.
11. Audio interview conducted by the author in 1975. Beginning in 1924, NAA, the American time signal station in Arlington, Virginia, began sending audio transmissions on the new AM broadcast band. Michael A. Lombardi, "Time Signal Stations," http://tf.boulder.nist.gov/general/pdf/2131.pdf, 125.
12. Walton, "Cap Cornish."
13. Shawn Gary VanCour, "The Sounds of 'Radio': Aesthetic Formations of 1920s American Broadcasting" (Ph.D. diss., University of Wisconsin–Madison, 2008), 2.
14. "KDKA Begins to Broadcast," A Science Odyssey: People and Discoveries—PBS, http://www.pbs.org/wgbh/aso/databank/entries/dt20ra.html.
15. "Partnership Is Dissolved: Clarence F. Cornish Now Sole Owner of Wayne Radio Co.," *News-Sentinel*, October 21, 1922, 3.
16. Unidentified newspaper clipping.
17. Cornish quote from 1975 audio interview with author. This is presumably a reference to Edwin Howard Armstrong, a prolific inventor who who fought many radio-related patent battles. Widely known as "the father of FM radio," he was awarded the Franklin Medal, the highest honor in U.S. science, in 1941. He was granted forty-two patents during his life, including U.S. Patent No. 1,424,065, which was granted on July 25, 1922, for "Signaling system (superregenerative receiving circuit for electron tubes)."
18. "Who's Who," *Journal-Gazette*, March 2, 1924.
19. "George W. Carter Speaks: Former Local High School Teacher Talks on Radio," unidentified newspaper clipping.
20. The history is printed on a single sheet of paper. Author's collection.
21. Cornish, Pilot's Book.
22. With rare exceptions, he continued to attend annual two-week training camps, at locations such as Wright Field and Schoen Field at Fort Benjamin Harrison in Indianapolis, until he was called to serve in World War II.

23. "Aviation in Fort Wayne," *Journal-Gazette*, January 18, 1925, 1.

24. "Huge Crowds Watch Airplane Flights: 40 Enlist in Flight B Unit; Will Take Oath Tuesday Night," clipping from the *News-Sentinel*, undated, but probably September 24 or 25, 1924.

25. "Their Stunts in Army Planes Here Sunday Aroused Interest in Reserve Recruiting," *News-Sentinel*, September 25, 1924.

26. Fort Wayne native Paul Frank Baer was an air ace in World War I. He enlisted in the Franco-American Flying Corps (the famous Lafayette Escadrille) in 1917 and transferred to the U.S. Flying Corps in 1918. He is officially credited with bringing down nine German planes before his capture late in the war.

27. Handwritten note; undated. Author's collection.

28. "Best Aviation Field in State Proposed," unidentified newspaper clipping.

29. "Crowd Expected for Defense Day Flying Program," *Journal-Gazette*, June 25, 1925, 1.

30. "Noted Flyers Here for Dedication Program at Baer Field Yesterday," *News-Sentinel*, June 26, 1925.

31. Early on, according to author William Thumma, Baer Field was sometimes referred to as Pennell Field. Thumma, *Early Aviation in Indiana* (Elwood, Ind.: W. Thumma, 1989), 40. That may have been in reference to John C. Pennell, the president of the Pennell Auto Company in Fort Wayne, who owned several planes and believed that there was "a great possibility in the passenger-carrying and advertising end of aviation." "The Aircraft Trade Review," *Aerial Age Weekly*, August 30, 1920, 834.

32. "Defense Stressed as City Dedicates Paul Baer Field: Mayor Hosey Tenders Site to the Cause of Aviation; 5,000 Gather to Witness Ceremonies," *Journal-Gazette*, June 25, 1925, 1.

33. Informal audio interview conducted by William W. Barrett at the home of Betty Nicholas, March 4, 1995. The Curtiss Oriole was made of laminated wood, accommodated a pilot and two passengers, and had a tall, thin radiator in the pilot's field of view.

34. Paul Hobrock also designed and built the Sky Romer, a four-seat high-wing cabin monoplane. For details, see "The Sky Romer," *Vintage Airplane*, January 1982, 12–14; and http://1000aircraftphotos.com/Contributions/Shumaker/10593.htm. In 1937, the Sweebrock Aviation Co., based at Paul Baer Municipal Airport, received a contract to supply the U.S. military with 14,000 towing targets for aerial practice.

35. Commercial flights would move from Guy Means Field (which Cornish was said to have managed in the late 1920s) to Paul Baer Field in 1931. Thumma, *Early Aviation in Indiana*, 39, 40.

36. Unidentified newspaper clipping.

37. Clarence F. Cornish Interview, Douglas E. Clanin World War II Oral History Collection, 1944–2002, Collection M 0783, Manuscript and Visual Collections Department, William Henry Smith Memorial Library, Indiana Historical Society, Indianapolis

[hereafter IHS interview]. Clanin was an editor with the Indiana Historical Society. Between 1983 and 2002, he conducted interviews with World War II veterans. The Clanin collection includes both the original audiotapes of the interviews and later transcriptions; the transcripts of Cornish's interview, which was conducted on January 11, 1992, are in box 4, folders 11–12.

38. Finis Farr, *Rickenbacker's Luck: An American Life* (Boston: Houghton Mifflin, 1979), 106, 107, 105.

39. Barrett interview.

40. Walton, "Cap Cornish," 12–13.

41. Donald B. Holmes, *Airmail: An Illustrated History, 1793–1981* (New York: Clarkson N. Potter, 1981), 158.

42. IHS interview.

43. "Accused Man Waives Extradition to Return to City in Plane with Detective," *News-Sentinel*, August 22, 1925, 1.

44. "The Greatest Array of Immense Commercial Airplanes Ever Assembled," unidentified newspaper clipping.

45. "Fort Wayne and Flying," *Journal-Gazette*, September 28, 1925, 4.

46. Officers of the Fort Wayne chapter of the National Aeronautic Association made the local arrangements for this tour: president, Arthur F. Hall; vice-president, Robert M. Feustel; secretary, F. H. George; treasurer, A. A. Burry; and general chairman, Robert R. Bartel. Cornish acted as "referee."

47. "Ford Tour Planes Arrive in Chicago: First and Second Legs of Journey Completed without Serious Mishaps; Praise for Fort Wayne," *Journal-Gazette*, September 29, 1925, 1.

48. "Giant Air Liners of German and American Make in Ford Air Tour and One Pilot and His Wife," *Journal-Gazette*, September 29, 1925, 6.

49. "Ford Tour Planes Arrive in Chicago."

50. Fort Wayne's geographic location was considered fortuitous by some. "City Located in Air Lane Usually Safe from Storms: Community's History Shows It Always Played Important Part in Nation's Transportation Progress—Indiana Sought It for Travel, Earlier Settlers Recognized It as Splendid Railroad Route—Now Air Course," *News-Sentinel*, July 28, 1928, 24.

51. Manufacturers Aircraft Association, *Aircraft Year Book* (New York: Manufacturers Aircraft Association, 1919), 32, 37, http://archive.org/stream/aerospaceyearbo00amergoog#page/n12/mode/2up; Aerospace Industries Association, "About AIA: History," http://www.aia-aerospace.org/about_aia/aia_at_a_glance/history/.

52. "Airplane Carries News-Sentinel and Mail to Fort Wayne Men Making Tour," clipping from the *News-Sentinel*, undated, but mid-October 1925, pt. 2.

53. "News-Sentinel to Go by Plane," clipping from the *News-Sentinel*, undated, but mid-October 1925.

54. "Young Matron Takes Plane Ride in Effort to Recover Hearing," and "Drops in Plane 12,000 feet; Deafness Cured," undated clippings from the *Journal-Gazette*.

55. A. Scott Berg, *Lindbergh* (New York: G. P. Putnam's Sons, 1998), 84.

56. Author's collection.

57. "Airplane-Radio Experiment Here Proves Success," *Kokomo Tribune*, January 2, 1926, 8.

58. "Making It Safe for Pilots," ibid.

59. "Thousands Thrilled by Air Races, Stunt Flying Here Sunday—Everybody Pleased with Two Day Program Carried Out at Sharp Field," Peru *Journal-Chronicle*, July 25, 1927, 1.

60. Ibid.

61. *News-Sentinel*, August 8, 1927, 6.

62. "Thousands See Air Circus," *Journal-Gazette*, August 28, 1927, 1.

63. When the American Legion held its state convention in Fort Wayne in 1930, the organization sponsored another air circus, during which Cornish won two events, one being a twenty-one-mile closed course that he completed in thirteen minutes, one second. "Flying Carnival Awes Thousands," *News-Sentinel*, August 25, 1930, 1.

64. Lois Watterson Cornish maintained a diary for many years; information from some of the entries is included here. Author's collection.

65. From letters written by Clarence to Lois in the 1920s. Author's collection.

66. The following details about the Cornishes' personal and professional lives are taken from the 1975 audio interview with the author.

67. "Form Corporations," unidentified newspaper clipping. According to a newspaper ad, the Independent Supply Company was a distributor for Kellogg, Howard, Premier, Shamrock, Dublier, Pacent, All American, U.S. Tool, Amsco, Elgin, Magnavox, Eiseman, Atlas, and Burgess Batteries. "Who's Who," *Journal-Gazette*, March 2, 1924, 7.

CHAPTER 4

1. Walter J. Boyce, *The Smithsonian Book of Flight* (Washington, D.C.: Smithsonian Books, 1987), 116.

2. The Auto Electric and Radio Equipment Company was headed by Matt Jones, Ivan Hitchcock, Arthur F. Hall, and Benjamin F. Geyer. It was later sold to Hall's son William B. F. Hall, to whom Cornish gave basic flying instructions and who later received a commission as a Navy pilot. From Clarence F. Cornish Interview, Douglas E. Clanin World War II Oral History Collection, 1944–2002, Collection M 0783, Manuscript and Visual Collections Department, William Henry Smith Memorial Library, Indiana Historical Society, Indianapolis [hereafter IHS interview]. Clanin was an editor with the Indiana Historical Society. Between 1983 and 2002, he conducted interviews with World War II veterans. The Clanin collection includes both the original audiotapes of the interviews

and later transcriptions; the transcripts of Cornish's interview, which was conducted on January 11, 1992, are in box 4, folders 11–12.

3. Advertisement, "Our Eyes Turn to the Sky!," *News-Sentinel,* March 8, 1928, 2.
4. "Boost Fort Wayne Aviation," *News-Sentinel,* July 2, 1928.
5. Ibid.
6. Cornish, Pilot's Book.
7. Unidentified newspaper clipping.
8. Jimmy Doolittle is credited with the development of the necessary instruments for blind flying. He made his first public demonstration under the hood at Mitchell Field, Long Island, New York, on September 24, 1929, when he successfully took off, flew fifteen miles, and landed his Consolidated airplane. Welman A. Shrader, *Fifty Years of Flight: A Chronicle of the Aviation Industry in America, 1903–1953* (Cleveland, Ohio: Eaton Manufacturing Co., 1953), 44. The final page in Cornish's Pilot's Book lists four dates for "Blind Flying in Fleet with K-5 engine" and five dates in 1936 and 1937 in a "'BT 2B' with Wasp engine." Two of the latter included cross-country flights to Detroit and Indianapolis.
9. Assen Jordanoff, "Blind Flying," *Popular Science,* November 1930, 39.
10. "Open Hangar," undated clipping from the *News-Sentinel.*
11. "Fort Wayne Pilots Win Prizes at Aviation Meet," unidentified clipping dated September 4.
12. Author's collection.
13. "Santa Claus Adopts Twentieth Century Method of Transportation in Trip Here from Frozen North," *News-Sentinel,* December 24, 1928.
14. Bob Schott, "Open Hangar," *News-Sentinel,* October 22, 1932.
15. "Patterson-Fletcher Co. Forms . . . Club," *News-Sentinel,* March 24, 1931, 27; "Flying Club Plane Arrives: Pusher Type Bought for Patterson-Fletcher Project," *News-Sentinel,* March 30, 1931, 19.
16. Unidentified newspaper clipping.
17. Ibid.
18. Miscellaneous unidentified newspaper clippings in author's collection.
19. In the early 1930s, Department of Commerce air regulations required fifty hours of solo flight before someone could be licensed as a private pilot with permission to carry passengers. Statistics show that 83 percent of the aviation accidents in 1932 involved pilots with less than fifty hours of solo flying.
20. "Club to Hear War Flier: Lieut. Cornish to Address Men of Temple Thursday," unidentified newspaper clipping.
21. In November 1929, Eielson crashed his plane in a blizzard while attempting to evacuate furs and personnel from a cargo vessel trapped in ice off the coast of Siberia. He and his mechanic, Earl Borland, were both killed. "Carl Ben Eielson: Aviation Pioneer," United States Air Force Fact Sheet, http://www.eielson.af.mil/shared/media/document/AFD-061114-009.pdf.

22. Unidentified newspaper clipping. On August 15, 1935, Post's aircraft crashed in a lagoon near Point Barrow, Alaska, after the motor failed. Both he and the famous American humorist Will Rogers were killed. The two were friends and had been enjoying a "happy-go-lucky aerial tour of Alaska" when the accident occurred. "Both Houses of Congress suspended deliberations upon learning of Rogers's and Post's deaths, and a period of national mourning followed with messages of condolence issued by leaders and governments throughout the world." National Park Service, Aviation: From Sand Dunes to Sonic Booms, "Rogers-Post Site," http://www.nps.gov/nr/travel/aviation/rps.htm.

23. "Thousands See Autogiro Here," *News-Sentinel*, August 24, 1931, 13.

24. "Mussolini Is Host to American Fliers," *New York Times*, July 12, 1929; "Italy Gives Medals to the Rome Fliers," *New York Times*, July 13, 1929.

25. "Thousands See Autogiro Here."

26. T. A. Heppenheimer, "When's the Last Time You Caught a Ride in an Autogiro?" *Air and Space*, March 2003, http://www.airspacemag.com/history-of-flight/cit-heppenheimer.html; Virginia P. Dawson and Mark D. Bowles, eds., *Realizing the Dream of Flight: Biographical Essays in Honor of the Centennial of Flight, 1903–2003* (Washington, D.C.: National Aeronautics and Space Administration, 2005), 76.

27. Shrader, *Fifty Years of Flight*, 49.

28. Text of speech in author's collection.

29. "History of Women Pilots: History of the 99s," Museum of Women Pilots, http://www.museumofwomenpilots.com/Histof99.html. Dorothy Mulligan, a swimming instructor at the YWCA, was the first girl in Fort Wayne girl to take a flying course; Ilo McCoy of Columbia City was the second. Unidentified newspaper clipping.

30. Author's collection.

31. Lloyd B. Walton, "Cap Cornish—On a 'High' That's Lasted since 1918," *Indianapolis Star Magazine*, October 7, 1973, 10.

32. "Hundreds See Air Program: Talk by Major Doolittle, Dead-Stick Stunting by Captain Cornish Feature Show at Paul Baer Field," *News-Sentinel*, July 9, 1934, pt. 2. Cornish was commissioned captain on June 7, 1930.

33. Harry Golding, *The Wonder Book of Aircraft for Boys and Girls* (London: Ward, Lock, 1919), 89.

34. Personal communication, 1996.

35. Others, however, were still eager to make a purchase from Aereco, including W. B. Standz of Lagrange, Indiana, who bought a Waco 10 with a Siemens-Halske engine. "Aereco Sells Plane," unidentified newspaper clipping.

36. "Cornish Sets New Unofficial Plane Efficiency Record," *News-Sentinel*, March 14, 1930, 1.

37. "Fort Wayne-Detroit Flight Made in 43 Minutes by Cornish," unidentified newspaper clipping. This flight is confirmed in Cornish's Pilot's Book.

38. "Cornish Flies to St. Louis in Small Plane for $4.55," unidentified newspaper clipping. This flight is confirmed in Cornish's Pilot's Book.

39. IHS interview.

40. Edmund F. Ball, *Rambling Recollections of Flying and Flyers* (Muncie, Ind.: Edmund F. Ball, 1993), 58.

41. Letter from Cornish to Mrs. Fran Wright, Sorrento, Fla., February 29, 1980. Author's collection.

42. Iwan Morgan, "Fort Wayne and the Great Depression: The Early Years, 1929–1933," *Indiana Magazine of History* 80, no. 2 (1984): 130.

43. "Nearly Score of Aviators Here in Benefit Air Show," *News-Sentinel*, September 16, 1932, pt. 2; "Air Show Is Successful: Record Crowd at Municipal Airport Sunday to Witness Event Staged for Relief Agencies' Benefit," *News-Sentinel*, September 19, 1932, pt. 2.

44. "2,500 Attend Airport Show: Indianapolis, Fort Wayne Airmen Stage Interesting Program for Benefit of Needy; Proceeds Low," *News-Sentinel*, October 23, 1933, pt. 2.

45. Their first home was a frame bungalow at 2414 Eastbrook Drive, which they bought for $3,800 in 1932.

46. Audio interview with author, 1975.

47. Ibid. Their payments were $19 a month.

48. For Oscar Foellinger and his family, community always transcended self-interest. Their legacy lives on through the Foellinger Foundation, the Foellinger Theatre in Franke Park, and the Foellinger-Freimann Botanical Conservatory in Fort Wayne, as well as Foellinger Auditorium at the University of Illinois, Urbana.

49. A wallet-sized card certifies that Mrs. Clarence Cornish was a passenger. Author's collection.

50. David Brinkley, *Washington Goes to War* (New York: A. A. Knopf, 1988), 76.

51. Steve Vogel, *The Pentagon: A History* (New York: Random House, 2007), 132.

52. "Glider Club Idea Approved: News-Sentinel Representatives Return from St. Louis Show after Conferences," *News-Sentinel*, February 24, 1930, 1.

53. "'Lindiana' to Be Glider's Name," *News-Sentinel*, March 1, 1930, pt. 2.

54. "Gliding to Be Demonstrated: Public Exhibition to Be Staged Here Sunday Afternoon at Guy S. Means Airport," *News-Sentinel*, May 2, 1930, pt. 2; "Glider Club to Be Formed Soon: News-Sentinel Laying Plans Thoroughly for Novel Aeronautical Movement Here," *News-Sentinel*, February 27, 1930, 1.

55. "Judges Chosen for Glider Day," *News-Sentinel*, September 17, 1931, 21.

56. "Yankee Clipper Passenger List First Year, 1,472," *News-Sentinel*, November 8, 1930, 19.

57. Unidentified newspaper clippings with titles such as "First School Flights Taken" and "Winners to Get Rides in Clipper."

58. "Pupil Describes Beauties of 'Yankee Clipper' Trip," *News-Sentinel,* December 12, 1929, 1.

59. Unidentified newspaper clipping.

60. "Carriers Win Plane Flight," *News-Sentinel,* May 2, 1932, 18; "Group Visits Dayton Field," *News-Sentinel,* May 6, 1932, pt. 2.

61. "Party Returns after Air Trip: Four Men Who Undertook Journey to California Still Air-Minded in Spite of Mishap," *News-Sentinel,* March 31, 1931, pt. 2.

62. "Big Throng Is Attracted by Air Show: Crowd at Kroger Aviation Carnival Is Estimated at 20,000; Cornish Is the Victor in Race Feature," *News-Sentinel,* May 6, 1934, pt. 2; photo of Cornish holding trophy, *News-Sentinel,* May 7, 1934, 18.

63. Helene Foellinger "'was really quite the pioneer,' said Cheryl Taylor, president and CEO of the Foellinger Foundation. 'Helene had plenty working against her: She was young, she was a woman and it was the 1930s. But against all odds—at age 25—she became one of the first women publishers in the country.'" Chelsea Brune, "Helene Foellinger: Pioneer in the Industry," News-Sentinel.com, July 7, 2008, http://fwnextweb1.fortwayne.com/ns/projects/175anniv/0707anniv2.php/ (accessed May 13, 2013). See also Joseph F. Sheibley, "Helene Foellinger," Indiana Journalism Hall of Fame, http://indianajournalismhof.org/1974/01/helene-foellinger/.

64. There were ten beacons along the airway, positioned atop fifty-foot-tall steel towers placed 100 to 250 miles apart (the distance was smaller in the mountains, greater in the plains). Each had a sequential number between 1 and 10, with the sequence repeated in each new stretch. Every rotating beacon flashed a coded pattern based on its location within its particular stretch of airway. The Morse code mnemonic for the beacon order was When Undertaking Very Difficult Routes Keep Heading By Good Methods. Pilots could determine their location along an airway by the Morse code letter a beacon flashed.

65. Daniel L. Rust, *Flying across America: The Airline Passenger Experience* (Norman: University of Oklahoma Press, 2009), 37.

66. "Plan to Honor Smith's Memory: Inaugural Ceremony for Opening of New Air Mail Line Here Saturday Is Arranged," *News-Sentinel,* December 3, 1930, 1, 8.

67. Peter Garrison, "By Stars, Beacons, and Satellites: How We Got from Point A to Point B," *Air and Space,* March 2006, 50–57.

68. Michael C. Hawfield, "The North Side: The Ace and the Daredevil," in *Fort Wayne Cityscapes: Highlights of a Community's History* (Northridge, Calif.: Windsor Publications, 1988), 43–46.

69. "Mail Official to Be in City," *News-Sentinel,* December 4, 1930, 1.

70. "Air Mail Ships Now Operating: First Round Trip to South Bend on New Shuttle Line Completed This Morning; Visibility Still Poor," *News-Sentinel,* December 8, 1930, 1.

71. Hawfield, "North Side," 43.

72. Roger Myers, conversation with the author, 2010; Michael Hawfield, "Cityscapes Past," *News-Sentinel,* June 16, 1984.

73. "Explains Air Mail Trouble: Captain Cornish Tells Hi-Y Clubs Commercial, Army Pilots Both Have Their Own Niches to Fill," unidentified newspaper clipping.

74. "3 to Fly to Capital for Route Confab," *News-Sentinel*, May 22, 1934, pt. 2; "Air Mail Route Meets Refusal: P. O. Dept. Calls Local Plan Expensive and Unnecessary," *Journal-Gazette*, May 24, 1934, 1.

75. Cornish assumed management of the airport on August 1, 1934. Two months later, the *News-Sentinel* announced that Aereco had disposed of its equipment and was going out of business. "Airplanes Sold: Aereco Service Disposes of Equipment; Will Be Dissolved," *News-Sentinel*, October 16, 1934.

CHAPTER 5

1. Most flyers still were men, but after Amelia Earhart accompanied pilot Wilmer Stultz and mechanic Louis Gordon on a flight from Newfoundland to Wales in 1928, becoming the first woman to cross the Atlantic in a plane, the interest among women grew. "Amelia was a strong advocate of awakening women's potential. She encouraged young girls to dream big, and said about women and aviation: 'The more women fly, the more who become pilots, the quicker will we be recognized as an important factor in aviation.'" "History of Women Pilots: History of the 99s," Museum of Women Pilots, http://www.museumofwomenpilots.com/Histof99.html.

2. "Aviation Group to Meet on Wednesday," *News-Sentinel*, September 18, 1933, 13; "Aviation Club Will Disband after Party," *News-Sentinel*, September 21, 1933, 3.

3. "New Flying Club Changes Its Name," unidentified newspaper clipping.

4. "Chicago Aero Club Girls at Airport," *News-Sentinel*, July 28, 1934, 4.

5. "Amby Babbitt Gets Low Score in Aerial Golf at City Airport," *News-Sentinel*, August 1934, 12.

6. "Ten Planes Leave City," *News-Sentinel*, October 12, 1935, 24; "Air Cruise under Way," *News-Sentinel-Gazette*, August 22, 1936, 1; "Eleven Planes Enter Cruise: Journal-Gazette Safety Trophy on Display," *Journal-Gazette*, August 20, 1936, sec. 2; "Wins Journal-Gazette Safety Trophy," *Journal-Gazette*, August 23, 1936, 1.

7. "Record Crowd Sees Air Show Here Sunday," *Journal-Gazette*, July 14, 1935, 1.

8. "Aero Club Series Set: First of Group of Educational Programs to Be Given at C. of C. Free of Charge on Wednesday," unidentified newspaper clipping.

9. "Taking Us for a Ride," editorial cartoon, *News-Sentinel*, July 14, 1939.

10. "Blaze Destroys Tri-Motor Air Craft at Show: $11,500 Display Plane of Pilot Murphy Burns as 20,000 See Stunts," *News-Sentinel*, July 17, 1939, 1, photo on 11.

11. "Airport Fire Truck Planned by Aero Club," *News-Sentinel*, January 10, 1940, 4.

12. Fort Wayne Aero Club minutes, December 19, 1939.

13. From Cornish's handwritten notes.

14. William Thumma, *Early Aviation in Indiana* (Elwood, Ind.: W. Thumma, 1989), 20.

15. Sturm was an Indianapolis newspaperman. In 1916 he accompanied fellow Hoosier E. G. "Cannonball" Baker on a cross-country driving trip as the latter set a transcontinental record in "a standard eight-cylinder Cadillac roadster," covering the 3,471 miles from New York to Los Angeles in eleven days, seven hours, and fifteen minutes. "Auto Makes Record in Run from Pacific," *New York Times*, May 16, 1916.

16. "Flying Vastly Different Than Several Years Ago," *News-Sentinel*, September 14, 1929, 19.

17. Itinerary map, ibid.

18. "First All-Indiana Air Caravan to Be Here Tuesday," *News-Sentinel*, September 14, 1929, 19.

19. "Air Tour Near City," *News-Sentinel*, September 17, 1929, 1.

20. According to an article titled "WPA to Mark Airway Signs: Allotment Granted for Painting Arrows," *Journal-Gazette*, February 8, 1938, the CWA had launched a statewide program, which was then interrupted when the agency became the WPA. With new funding in 1938, fourteen communities in Indiana would gain new markers; markers in thirty-six communities would be renewed.

21. "First All-Indiana Air Caravan to Be Here Tuesday."

22. "Air Tour near City."

23. Associated Press, Richmond, Ind., "New Stowaway Stunt," September 17, 1929.

24. "What Your Chamber of Commerce Is Doing Now," *Bulletin of the Indianapolis Chamber of Commerce* 45, no. 6 (June 1930): 1.

25. Prohibition was still in effect at the time, but the Berghoff Brewery had managed to stay in business "by producing 'Bergo,' a soft drink similar to root beer, and Berghoff Malt Tonic, which was said to be useful in aiding both the young and old in fighting a number of health problems. After Prohibition was repealed, the brewery was the 1st in Indiana to produce legal '3.2' beer." "Berghoff History," www.berghoffbeer.com/history/.

26. Mary Bostwick, "Forty Planes in Air Tour Back Home: Clarence Cornish of Fort Wayne Awarded Safety Trophy amid Cheers," *Indianapolis Star*, June 27, 1933.

27. Photo, "Women Pilots on Annual All-State Air Tour," *News-Sentinel*, June 24, 1933.

28. "25 Planes in Tenth Annual Indiana Tour," *News-Sentinel*, October 10, 1938, pt. 2; Ball, *Rambling Recollections*, 140; "Cornish Wins Tour Trophy: Fort Wayne Pilot Given Lincoln Life Co. Award for Safety Record on Indiana Air Journey," *News-Sentinel*, June 26, 1933, pt. 2; Bostwick, "Forty Planes in Air Tour Back Home."

29. "Air Tour Pilots Will Return Here Tuesday for Party," unidentified newspaper clipping; "Pilots of Air Tour Guests of Berghoff Firm at Party Here," unidentified newspaper clipping.

30. "Tourists Encounter Rain: Planes on Jaunt over State Show Signs of Storm Battles," unidentified newspaper clipping; "Air Tourists Fight Weather: Arrive at City Field Today," *Journal-Gazette*, September 14, 1934, 1.

31. "Planes Move to Wawasee," *News-Sentinel*, September 13, 1934, pt. 2.

32. "Air Tourists Fight Weather."

33. "Air Caravan Stops Here," *News-Sentinel*, September 15, 1934, pt. 2.

34. "Air Tourists Move On to End of Trip," *News-Sentinel*, September 15, 1934, pt. 2. Murphy would become a highly respected glider pilot in World War II.

35. Ibid.

36. "City Is Host to Indiana's 6th Air Tour," *Journal-Gazette*, September 15, 1934, 1.

37. "All-Indiana Air Tour June 17–23," *News-Sentinel*, February 22, 1935.

38. "Start Second Lap of State Air Tour," *News-Sentinel*, June 18, 1935, 24.

39. " Famous Flyer May Join Tour," *Journal-Gazette*, June 18, 1935, 20.

40. "Tour Planes Here Friday," *News-Sentinel*, June 19, 1935, pt. 2.

41. Unidentified newspaper clipping.

42. "Tour Planes Arrive Here: Advance Guard Followed by Flotilla of Forty Ships; Mayor and Civic Leaders Welcome Visitors to City," unidentified newspaper clipping.

43. Charles A. Keefer, "4,000 Welcome State Air Tour at Baer Field." *Journal-Gazette*, June 22, 1935, 1.

44. Although Cornish was dedicated to promoting the tour, serving on the IATA's tour committee under Herbert O. Fisher and as chair of Fort Wayne's advisory committee, he would not be an entrant that year because of major construction projects underway at the Paul Baer Municipal Airport.

45. According to his cousin Edmund F. Ball, the unmanageable aircraft simply fell out of the sky. Ball, *Rambling Recollections*, 47; "Frank Ball Killed," *News-Sentinel*, May 29, 1936, 1.

46. "Before leaving Muncie, Walker Winslow placed a wreath on Frank E. Ball's mausoleum in Beech Grove Cemetery; and departing, a formation of aircraft flew overhead, paying tribute to Frank for what he had contributed to aviation in his few short years as a pilot." Ball, *Rambling Recollections*, 140.

47. "Tour Enters Second Day," *News-Sentinel*, June 23, 1936, 28.

48. "City Ready to Welcome Air Armada," *News-Sentinel*, June 27, 1936, 1.

49. "Tour Enters Second Day."

50. "Announce Air Tour Route," *News-Sentinel*, June 19, 1936, 4.

51. "Air Tourists Have Narrow Escape in Crash," *News-Sentinel*, June 27, 1936, 1.

52. "Two Injured at Air Show," *News-Sentinel*, June 29, 1936, 8.

53. "Air Tour Parley Ends in Deadlock: Pilots Association to Proceed with Plans for Tour June 14–19," *News-Sentinel*, April 22, 1937, pt. 2.

54. Ball, *Rambling Recollections*, 139.

55. Bostwick was also known for having been the first woman to ride around the Indianapolis Motor Speedway's 2.5-mile track, in 1922.

56. Ball, *Rambling Recollections*, 139.

57. "25 Planes in Tenth Annual Indiana Tour."

58. "Cornish Heads Aircraft Group: Re-elected President of Indiana Organization; Plan to Sponsor Air Tour," *News-Sentinel*, February 19, 1940, 1.

59. The Capehart Corporation made radio receivers in Fort Wayne. It was founded by Homer Capehart, who later became a U.S. senator, serving from 1945 to 1963.

60. IHS interview with Douglas Clanin.

61. That was certainly the case for Cornish and fellow aviator Ed Ball. The two men remained friends for the rest of Cornish's life. They corresponded into the 1990s about the establishment of an Indiana air museum at Purdue University, and after Cornish's death, it was Ball who nominated him to be honored in the Aviation Hall of Fame.

62. "Cornish on Committee: Fort Wayne Man Chosen to Represent Aviation Interests of Indiana at Meeting in Washington," unidentified newspaper clipping.

63. "To Indianapolis: Pilots Plan Drive at Capital against Air Department," *Journal-Gazette*, January 17, 1935.

64. "State Commission on Aviation Likely: Cornish May Propose Establishment of a Non-Pay Council of Five," unidentified newspaper clipping. (Although this headline mentions five members, the proposal that subsequently was submitted called for eight.)

65. Ibid.

66. "Bill to Capital: Cornish-Helmke Aviation Commission Proposal Drawn Up," unidentified newspaper clipping.

67. "State Commission on Aviation Likely."

68. "State Air Tour Set for June 22: Cornish Named on Study Committee," *Journal-Gazette*, February 29, 1936.

69. The radio range was a navigation system that defined electronic airways by means of a network of radio beacons, which transmitted signals in Morse code every thirty seconds to help pilots orient themselves. "If the aircraft drifted off course to one side, the Morse code for the letter 'A' could be faintly heard. The greater the drift, the stronger the 'A' Morse code signal. Straying to the opposite side produced the 'N' Morse code signal." But "when the aircraft was centered on the airway, or electrical beam, these two opposite Morse-code signals merged into a steady, monotonous, hypnotizing tone." Thus the common expression "on the beam," in the sense of doing something correctly or heading in the right direction. Richard L. Taylor, *Forty-Seven Years in Aviation: A Memoir*, chap. 7, pt. 3, http://www.avweb.com/news/skywrite/forty_seven_years_chap_7_205599–1.html.

Chapter 6

1. "Cornish to Be on Program at Aircraft Show: Will Discuss Model Airports at International Event in Chicago," *News-Sentinel*, January 19, 1938, pt. 2; "Model Airports Not at All Like Topsy, Says Capt. Cornish," *News-Sentinel*, February 3, 1938, pt. 2.

2. "Cornish Named Airport Head," *Journal-Gazette*, July 11, 1934, 1.

3. "Bartel Ousted as Municipal Port Manager," *Journal-Gazette*, June 30, 1934, 1.
4. Editorial, *News-Sentinel*, July 12, 1934.
5. "The Case of Capt. Bartel," *Journal-Gazette*, July 1, 1934, 4.
6. "Airport Development Planned," *News-Sentinel*, January 3, 1934, 6.
7. "Lower Budget Proposed for Airport Here," *News-Sentinel*, July 9, 1935, pt. 2.
8. "Plan Changes at Airport," *News-Sentinel*, August 16, 1934, pt. 2.
9. "Nose-Dives," *Journal-Gazette*, December 23, 1934.
10. "Plan Hockey Game," unidentified newspaper clipping.
11. "Curtis Publishing Company issued a report in 1930 that included suggestions for drawing more women to air travel. The report recommended that airports install pleasant outdoor landscaping and create tidy waiting rooms and attractive restaurants." Rust, *Flying across America*, 60.
12. The Rounder, "Around the Town," *Journal-Gazette*, October 8, 1934, 9.
13. "Airport Plans Recreations," *News-Sentinel*, July 20, 1933, 1; "Beer Sale to Aid Airport," *News-Sentinel*, July 25, 1933, pt. 2.
14. "Airport May Be Annexed to City by Leasing One-Foot Strip Two Miles in Length," *News-Sentinel*, October 29, 1935, 1.
15. Unidentified newspaper clipping.
16. FERA operated from May 1933 through December 1935 to alleviate household unemployment and develop facilities on public land. It was replaced by the WPA in 1935.
17. "Airport Paving Authorized," *Journal-Gazette*, July 17, 1935, 2.
18. "Increased Air Traffic Seen," *News-Sentinel*, July 25, 1935, pt. 2.
19. Rust, *Flying Across America*, 99.
20. Ibid.
21. "3 to Fly to Capital for Route Confab," *News-Sentinel*, May 23, 1934, pt. 2. The Thompson Aeronautical Corporation was reportedly the first to introduce airmail service to Fort Wayne, on December 6, 1930. In May 1931, TWA initiated Chicago to Columbus, Ohio, passenger service through the the city, adding mail service on its line on May 1, 1933. "City's Airport Ranks High Nationally for Modern Equipment," clipping from the *News-Sentinel*, undated, but around July 1, 1936.
22. "Many Planes Stop," *News-Sentinel*, July 26, 1935, 10.
23. "Eight Cruise Planes Here," *News-Sentinel*, September 21, 1935, 1.
24. "Air Mail Still a Possibility," *News-Sentinel*, December 13, 1935, pt. 2.
25. "Chances for Air Mail Hit," *News-Sentinel*, April 6, 1936, pt. 2.
26. Photo, *News-Sentinel*, March 29, 1935, 36.
27. Clifford B. Ward wrote several Abracadabra columns for the *News-Sentinel* each week in the 1930s. This quote is from an undated clipping.
28. "Local Air Officials Hail Famous Balloonist," *Journal-Gazette*, March 17, 1935, 4; "Race Offer Is Favored," *News-Sentinel*, March 18. 1935, pt. 2.

29. "Increased Air Traffic Seen: TWA Will Install Radio Set for Emergency Use," *News-Sentinel*, July 25, 1935, pt. 2. "By [1930], Cleveland Municipal Airport had established radio control of airport traffic. In the next five years approximately 20 cities followed Cleveland's lead." "FAA Historical Chronology, 1926–1996," http://www.faa.gov/about/media/b-chron.pdf.

30. "More Airport Funds Asked: Indiana's WPA Grant May Be Doubled, Major Cox Says; Municipal Airport Would Share in Increase," unidentified newspaper clipping.

31. As quoted in *Air Commerce Bulletin* 7 (1935): 75.

32. "Plans Ready for $278,000 WPA Airport Project," *News-Sentinel*, November 7, 1935, pt. 2.

33. "WPA Airport Plans Ready," *News-Sentinel*, November 7, 1935, pt. 2.

34. "Project to Go Forward," *News-Sentinel*, November 12, 1935, 4.

35. "Aviation Activities Greatest in Years," *News-Sentinel*, January 21, 1936, 8.

36. "Dream of Best Airport in State Nears Realization," *News-Sentinel*, April 2, 1936, 9. Technical issues had reversed the initial approval for some of these improvements, but the problems were resolved by the end of 1935. "$50,304 for Airport Work Is Approved: Revised Plans for Radio Range Beacon, Boundary Lights O. K.'d By Bureau of Air Commerce," *News-Sentinel*, December 26, 1935, pt. 2.

37. "Radio Range Beacon Built at Municipal Airport," *News-Sentinel*, May 30, 1939, 24. In the fall of 1939, a replacement radio range beacon providing four directional beams was installed on Carroll Road. "Federal Radio Range Beacon Nears Completion," *News-Sentinel*, November 15, 1939, pt. 2.

38. "Airport Boundary Lights Dedicated: Four Thousand Persons Attend 'Open House' Ceremony at Municipal Field," *News-Sentinel*, June 24, 1936, pt. 2.

39. "Ceremony at Airport Set and Airport Boundary Lights Dedicated," clipping from the *News-Sentinel*, undated, but around June 24, 1936, 1.

40. "$24,000 for Airport Work," *News-Sentinel*, May 14, 1936, pt. 2.

41. A decade later, in October 1946, W. L. Hempelmann, an engineer with the Texas Company's Asphalt Sales Department in Chicago, sent Cornish a meticulously hand-printed letter asking whether he "had changed after ridding himself of the nerve wracking details of paving the airfield." He closed with, "Please pardon this seeming familiarity but I harbor only the kindest of thoughts for your treatment 10 years ago of two asphalt representatives in the persons of A. J. Moynihan and W. L. Hempelmann." Enclosed was a photograph of Cornish standing in front of a boxy four-door sedan.

42. Numerous articles in both the *News-Sentinel* and the *Journal-Gazette*, February 2–6, 1937.

43. "TWA to Start Local Air Service Soon: U.S. Lets Mail Pact to Air Line Company," *Journal-Gazette*, July 30, 1937, 1.

44. "Plan to Start TWA Service within 10 Days," *Journal-Gazette*, August 3, 1937, sec. 2.

45. "Airline May Be Delayed," *News-Sentinel*, August 9, 1937, 14.

46. "Details of TWA Survey Flight Wednesday over Dayton–Fort Wayne Route Are Announced: Collings to Pilot Technical Ship; Safety Aids Will Be Demonstrated," *News-Sentinel*, August 19, 1937, pt. 2.

47. "Suggest City Sell Airport," *Journal-Gazette*, August 18, 1937. Airport expenses at the time were running around $20,000 a year, with income from operations only about $5,000.

48. "Airport Sale Hit by Mayor," *News-Sentinel*, August 18, 1937, pt. 2.

49. "Municipal Airport Development Program Crippled by Two-Mill Reduction in Next Year's Levy," *News-Sentinel*, September 16, 1937, pt. 2.

50. "Congress Is Asked for Big Increase in Naval and Airplane Strength," *News-Sentinel*, January 28, 1938, 1.

51. "Demands for Gigantic Air Force Voice: Bloc Is Forming in Congress to Insist on More Planes for U. S. Defense," *News-Sentinel*, February 26, 1938, 1.

52. "Hitler Makes Triumphant Entry into Austria," *News-Sentinel*, March 12, 1938, 1.

53. "National Attention Focused on Airport Here through Magazine," *News-Sentinel*, January 10, 1938, 17.

54. "WPA Airport Job Approved," *News-Sentinel*, March 28, 1938, pt. 2.

55. "U.S. Spends Big Sum Here," *News-Sentinel*, March 9, 1936, pt. 2.

56. "City, County Greatly Benefited by FERA, PWA during Past Year," *Journal-Gazette*, January 1, 1935, 9.

57 "National Airport Chief Praises Field," *News-Sentinel*, April 25, 1938, pt. 2.

58. "FAA Historical Chronology."

59. "Famed Transatlantic Flier to Be at Airport Here Four Days," *News-Sentinel*, October 18, 1938, 1.

60. "Col. Turner Says Fate of World Rests on Nations' Air Strength," *News-Sentinel*, October 21, 1938, pt. 2.

61. "City Will Hail TWA Opening Flight Tonight," *News-Sentinel*, December 15, 1938, pt. 2.

62. "TWA Connects City, Coasts in Night Service," *News-Sentinel*, December 16, 1938, 19.

63. *News-Sentinel*, January 12, 1939, 1, with full speech on 23.

64. "Cornish Named Region Leader," *News-Sentinel*, January 12, 1939, 4.

65. "Aviation Instruction for Boys to Be Recommended to Schools," *News-Sentinel*, February 25, 1939, 1, 9; "NAA Chapter Plans Ready," *News-Sentinel*, March 21, 1939, pt. 2.

66. "Speakers for Airport Forum May 5 Named," *News-Sentinel*, April 17, 1939, pt. 2.

67. "Airport Head Cites Need of Federal Funds," *News-Sentinel*, May 5, 1939, 1, 8.

68. "New Weather Service to Be Set Up Here: Government Approves Establishment of Airport Observation Branch," *News-Sentinel*, May 15, 1939, 1.

69. "Airport Wing for Weather Bureau Sought: $3,500 Needed for Addition to Administration Building for Units," *News-Sentinel*, April 28, 1939, pt. 2.

70. "Bids Sought at Airport," *News-Sentinel*, May 25, 1939, pt. 2.

71. "Venomous Reptiles and Insects in the London Zoo Killed," *News-Sentinel*, September 2, 1939, 1.

72. "Motion Pictures Planned in Schools: Open House Friday at the Airport," *News-Sentinel*, September 18, 1939, pt. 2.

73. "Air Highway System for 4 States Planned," *News-Sentinel*, October 28, 1939, 1. The participants included CAA officials and members of the Governor's Aviation Fact-Finding Committee, which Cornish chaired.

74. "Confab Is Set on Aviation," *News-Sentinel*, November 16, 1939, 36.

75. Charles Keefer, "Weather Bureaus, CAA Stations, like Silent Sentinels, Watch over Nation's Airways Travel," *News-Sentinel*, February 3, 1940, 3.

76. "New Weather Service to Be Set Up Here."

77. "Airport Now City-Owned: Bonds Paid; Final $20,500 Payment This Month Amortized Indebtedness," unidentified newspaper clipping.

78. "Home Building near Airport Cited as Peril," *News Sentinel*, February 7, 1940; "City Resists Move to Build near Airport," *News-Sentinel*, April 4, 1940, 32.

79. "City Is Denied Right to Block Airview Tract," *Journal-Gazette*, June 4, 1940, 1.

Chapter 7

1. This would include land west of the airport at the time to Lima Road between Cook Road on the north and Ludwig Road on the south, and all land lying south of Ludwig Road to Washington Center Road between the New York Central Railroad right-of-way on the east and Lima Road on the west. "Act to Add 330 More Acres to Baer Airfield," *Journal-Gazette*, August 21, 1940, 1.

2. "Nation Wide Air Traffic Survey Shows Paul Baer Field One of Busiest in U.S.," *Journal-Gazette*, August 26, 1940, sec. 2.

3. Brief Prepared for Submission to A. H. Wait, Regional Airport Engineer, by Lloyd S. Hartzler, Associate City Attorney, August 26, 1941.

4. "Army May Build Air Base Here," *Journal-Gazette*, August 14, 1940, 1.

5. "City May Lose Army Air Base," *Journal-Gazette*, December 10, 1940, 1.

6. "Civic Heads to Discuss Air Program," *News-Sentinel*, August 19, 1940, pt. 2.

7. "Act to Add 330 More Acres to Baer Airfield," 3. The approximate value of the acreage was $53,000, with a maximum projected cost to the city of $65,000.

8. "Air Program Group Named: Delegation to Washington Next Week," *Journal-Gazette*, August 27, 1940, sec. 2.

9. "Cornish Gives Confab Report: Aviation Committee to Push Project," *Journal-Gazette*, September 4, 1940, sec. 2.

10. "Group Leaves Tonight with Air Base Bid," *News-Sentinel*, September 4, 1940, pt. 2.

11. The site is now the location of the Fort Wayne International Airport, so named in 1991.

12. "Officers Inspect Proposed Air Base Site," *Journal-Gazette*, September 21, 1940, 1.

13. "Airport Here May Get Part of Big Fund," *News-Sentinel*, October 8, 1940, 1.

14. "Preside at Aviation Meeting," *Journal-Gazette*, October 2, 1940, sec. 2.

15. "Fort Wayne to Get Army Air Base," *News-Sentinel*, October 11, 1940, 1.

16. "Board Awaits More Data on U.S. Air Base: Airport Program Depends on More Detailed Information from War Dept.," *News-Sentinel*, October 12, 1940, 1.

17. Editorial, *Journal-Gazette*, October 13, 1940, 4.

18. "War Dept. Allots Millions for Army Construction Here," *News-Sentinel*, November 2, 1940, 1.

19. "Detail Plans for Air Base," *Journal-Gazette*, November 3, 1940, 1. Building plans then included twenty-seven barracks for enlisted men, seven day rooms, four enlisted men's messes, a cafeteria to serve one thousand persons, four officers' quarters, seven organization supply rooms, six administration buildings, a fire station, a guardhouse, a hospital to accommodate one hundred patients, an infirmary, a motor repair shop, a recreation building, a telephone building, a theater, two quartermaster warehouses, a post exchange, five operations buildings, a Link trainer building, a parachute building, a school building, a hangar, a radio station building, six storage buildings, and magazines. Work to prepare the stations was to begin "almost immediately."

20. "War Dept. Allots Millions for Army Construction Here."

21. "Aviation Base Here Looms More Likely," *News-Sentinel*, October 7, 1940, 1; "Fort Wayne to Get Army Air Base," *News-Sentinel*, October 11, 1940, 1.

22. "Air Official Coming Here," *News-Sentinel*, November 6, 1940.

23. "Air Corps Plans to Station 3,300 Men at Fort Wayne Base," *News-Sentinel*, November 3, 1940.

24. "Petitions Out for, against Air Base Here," *News-Sentinel*, November 22, 1940.

25. "First Steps Taken to Buy New City Port," *News-Sentinel*, November 27, 1940, 1.

26. Editorial, "Fort Wayne Gets the Air Base," *News-Sentinel*, December 12, 1940, 6.

27. "Meet Tuesday on Army Base: Will Hold Conference at Airport," *Journal-Gazette*, August 15, 1940, 12.

28. "Site of Proposed Army Base Southwest of City," *News-Sentinel*, December 11, 1940, 1.

29. Ibid.

30. "Army Base Acquisition Is Speeded," *News-Sentinel*, January 6, 1941, 1.

31. As of mid-1944, there were more than 250 buildings on the base, and more were underway. "The field is like a small city. It maintains its own water plant, which is fully automatic and produces 100 gallons per capita per day and can accommodate 10,000 persons. It has its own fire department, which has grown from one truck to a size comparable to that of a city with 10,000 population and includes a crash truck on duty 24 hours a day. The field has its own theaters, stores, heating facilities, police department, and so forth." "Baer Field," Extension of Remarks of Hon. George W. Gillie of Indiana in the U.S. House of Representatives, June 14, 1944, *Appendix to the Congressional Record*, 78th Cong., 2nd sess., vol. 90, pt. 5, A3012.

32. "Col. Lohman's Bomber Arrives," *Journal-Gazette*, April 13, 1941.

33. Brief Prepared for Submission to A. H. Wait, 6.

34. Ibid.

35. Cornish's personal files include a copy of the cover letter plus a six-page history and description of the situation at Smith Field as of August 2, 1942, sent by Mayor Baals to Lt. General H. H. Arnold, Chief of Army Air Forces.

36. "Effort Made to Keep 'Anthony Wayne' Name for Field despite New War Department Ruling," *Journal-Gazette*, April 13, 1941.

37. Hometown hero and World War I air ace Paul Frank Baer had been killed while flying mail in China on December 9, 1930. He is buried in Fort Wayne's Lindenwood Cemetery. "Nod of Approval for Name of New Air Field," *News-Sentinel*, June 9, 1941.

38. The monument was dedicated August 13, 1928. Michael Hawfield, "Airfields Named for Two Pioneers," *News-Sentinel*, June 16, 1984.

39. "Cornish to Suggest Naming New Airport in Honor of Smith," *Journal-Gazette*, July 11, 1941. Sometime after Smith's death and after contributions had been made to a memorial fund, Cornish wrote the editor of the *News-Sentinel* to support a suggestion by Fred George that an Art Smith Hangar be erected to house planes of transient pilots. An illuminated wind cone and a revolving light would be placed atop the hangar to "light the way of pilots traveling to and through the city at night." He continued, "It was in the air that Art Smith achieved his triumphs and glory and it is through intimate association and proximity of air craft activity that his memory would best be revered and perpetuated."

40. "May 6, 1978, The Art Smith Aeroplane Society honored Cornish on the [60th] anniversary of his first flight and stated on the certificate: His efforts, 1934-1941, caused Fort Wayne to have one of the finest ten airports in the nation; largely by an uncanny foresight, vast aviation experience and his wide knowledge of airway communications' technology. He was responsible for that facility being renamed to honor the memory of Art Smith." Signed by Ray Robinson, Secretary & Treasurer; Robert P McComb, President.

41. "Aero Club to Fly over City Friday Morning," *News-Sentinel*, July 24, 1941, pt. 2.

42. In September 1942, with the nation fully engaged in World War II, the club voted

to cease all further activities for the war's duration so that the members could redirect their energies to the war effort, which they aided by volunteering in the Civil Air Patrol. The Fort Wayne Aero Club resumed normal activities in January 1945, near the war's conclusion.

43. Author's conversation with Joe Marana, Operations Manager, Fort Wayne International Airport and Smith Field, January 28, 2014.

44. Author's conversation September 2009 with Tom O'Connell, Smith Field's Manager.

45. E-mail correspondence with Roger Myers, 2012.

46. Ibid.

47. "CAA Combines Hearings on Airline Stops: Fort Wayne Would Be Included on North-South Run on All Three Applications," *News-Sentinel*, June 20, 1941, pt. 2.

CHAPTER 8

1. "By direction of the President under the authority contained in Public Resolution No. 96, 76th Congress, approved 27 August, 1940, CAPTAIN CLARENCE FRANKLIN CORNISH (O-122897), Air Corps Reserve, is ordered to active duty, effective 5 July, 1941, with permanent station at Barksdale Field, Shreveport, Lousiana, for duty with the Air Corps Advanced Flying School." Quote from SO 144, 21 June, 1941, page 16, which listed men called to service. Author's collection.

2. Barksdale Field eventually became the headquarters of the "Mighty Eighth" Air Force and is home for the Air Force Global Strike Command.

3. Since Cornish's promotion from lieutenant to captain in 1930, he had been known to many as "Cap."

4. Wesley Frank Craven and James Lea Cate, *The Army Air Forces in World War II*, 7 vols., vol. 6: *Men and Planes* (Chicago: University of Chicago Press, 1955), 136.

5. "FAA Historical Chronology, 1926–1996," www.faa.gov/about/media/b-chron.pdf.

6. Clarence F. Cornish Interview, Douglas E. Clanin World War II Oral History Collection, 1944–2002, Collection M 0783, Manuscript and Visual Collections Department, William Henry Smith Memorial Library, Indiana Historical Society, Indianapolis [hereafter IHS interview]. Clanin was an editor with the Indiana Historical Society. Between 1983 and 2002, he conducted interviews with World War II veterans. The Clanin collection includes both the original audiotapes of the interviews and later transcriptions; the transcripts of Cornish's interview, which was conducted on January 11, 1992, are in box 4, folders 11–12. When Cornish began his assignment in November 1941, the Army Air Forces was a new entity. "Successor to the Army Air Corps and forerunner of the United States Air Force," it "owed its designation to Army Regulation 95-5 of 20 June 1941." Craven and Cate, *Army Air Forces in World War II*, 6:28. General Henry

H. "Hap" Arnold, who had been instrumental in the reorganization effort, was placed in command. The U.S. Air Force did not come into being until September 18, 1947, when, "based on the AAF's wartime achievements and future potential," it "won its independence as a full partner with the Army and the Navy." U.S. Air Force, "History," http://www.airforce.com/learn-about/history/part2/.

7. Letter dated December 4, 1941. Author's collection.

8. In a letter to Lt. Col. J. B. Hartranft, IATCB Secretary, dated September 13, 1945, Cornish wrote: "I gather from your letter [of September 3, 1945] that the IATCB is to carry on as a post-war coordinating agency. Is that true? If not, what seems to be the thoughts of the powers that be as to its future? I had always believed that there should be some sort of a body to coordinate the activities of the various agencies even though most of the policies based on wartime necessity could and should be eliminated as soon as possible." Author's collection.

9. Civil airways were created in 1926 to maintain safety during flight. A *Civil Airways of the United States and Canada* map dated June 1941 defines flight levels as follows: "Green and red civil airways: eastbound—odd thousand foot levels; westbound—even thousand foot levels. Amber and blue civil airways: northbound—odd thousand foot levels; southbound—even thousand foot levels. On display October 6, 2013, at the Howard W. Cannon Aviation Museum, McCarran International Airport, Las Vegas, Nevada.

10. Craven and Cate, *Army Air Forces in World War II*, vol. 1: *Plans and Early Operations, January 1939 to August 1942* (Chicago: University of Chicago Press, 1947), 152, 153, 271.

11. Gail Wells, "Unions and Hard Times between the Wars: World War II," Oregon History Project, ohs.org/education/oregonhistory/narratives/subtopic.cfm?subtopic_ID=574; "Investigations: Japanese Balloon Bomb," PBS: History Detectives, http://www.pbs.org/opb/historydetectives/investigation/japenese-balloon-bomb/; Johnna Rizzo, "Japan's Secret WWII Weapon: Balloon Bombs," *National Geographic Daily News*, May 27, 2013, http://news.nationalgeographic.com/news/2013/05/130527-map-video-balloon-bomb-wwii-japanese-air-current-jet-stream/.

12. "The High Cost of WWII," USS Hancock CV/CVA-19 Association, http://www.usshancockassociation.org/history-05.php; Second World War History: World War 2 Statistics, secondworldwarhistory.com/world-war-2-statistics.asp.

13. Wells, "Unions and Hard Times between the Wars."

14. Letter dated August 17, 1942. Author's collection.

15. "Air Space Reservations and Flight Hazards," in U.S. Army Air Forces, *Pilot's Information File* (Washington, D.C.: Army Air Forces, 1944–45), PIF 2-5-1.

16. Craven and Cate, *Army Air Forces in World War II*, 6:viii–ix.

17. On the night of January 16, 1942, TWA Flight 3 from New York City to Los Angeles carried actress Carole Lombard and others to their deaths. After making a

scheduled stop in Indianapolis, where it picked up Lombard and some other passengers, it flew on to Las Vegas, where it made a quick stop. "At 7:07 p.m., the flight departed runway 34 and began its climbing left turn across the Las Vegas Valley. The night of January 16th was dark and moonless as the DC-3 leveled off at the cruising altitude of 8,000 feet. The night was made even darker with the government's decision to blackout the lighted airway beacons due to wartime national security threats. . . . Neither pilot noticed the selected course was sending them into the snow-capped 8,500 foot Potosi Mountain." "January 16, 1942, Transcontinental & Western Air (TWA), Douglas DC-3 (NC1946) Potosi Mountain, NV," Lost Flights: Historical Aviation Studies and Research, http://www.lostflights.com/Commercial-Aviation/11642-TWA-TWA-Douglas-DC-3/5007934_mqCvFr.

18. True lights, as opposed to "false" lights, served as "true aids to aerial navigation." There were four general types: airport beacons, private airway beacons, landmark beacons, and hazard warning beacons. Arthur R. Nilson, "Radio Tower Lighting and Marking," http://www.americanradiohistory.com/Archive-Communications-Magazine/Communications%201937%2011%20November.pdf, 13.

19. IHS interview.

20. The request for the area above the St. Louis Ordnance Plant and Proving Grounds, at Goodfellow and Bircher Boulevards, to be designated as a danger area had been approved by the IATCB on December 5, 1941.

21. Airways at that time operated under the navigational principle of low frequency ranges, or LFRs, and were assigned colors that civilian pilots used to navigate from point to point. "For instance," Dick Cunningham explains, "you might follow Green Airway No. 4 to Amber Airway No. 7 from Indianapolis to Fort Wayne. Radio signals, transmitted from a set of five towers (one central and four legs), served as audio guides. If you strayed off course, the radio signal from a leg or line would become unbalanced, the sound would change in your headphones, and you would know to turn back. It was hard to do, especially during thunderstorms." Personal communication.

22. Military officials at Fort Knox warned that "no control over the angle of fire could be maintained due to uneven terrain." IATB minutes, November 12, 1941.

23. Ibid., September 9, 1941.

24. "Air Space Reservations and Flight Hazards."

25. IATCB minutes, September 9 and December 5, 1941.

26. Barrage Balloons in World War II: From Camp Tyson to Omaha Beach, http://www.skylighters.org/barrageballoons/; Elizabeth D. Schafer, "Barrage Balloons," in *World War II at Sea: An Encyclopedia*, ed. Spencer C. Tucker, 2 vols. (Santa Barbara, Calif.: ABC-Clio, 2011), 1:88–89.

27. IATCB minutes, April 17, 1942.

28. *The Officers' Guide: A Ready Reference on Customs and Correct Procedures Which Pertain*

to Commissioned Officers of the Army of the United States (Harrisburg, Pa.: Military Service Publishing Co., 1942), 270.

CHAPTER 9

1. The letters cited in this chapter are from the author's collection.

2. There were plenty of Stearmans around, since nearly ten thousand PT-17 Stearman (Boeing) Trainers were built in the 1930s and 1940s. Also known as Kaydets, they were the primary trainers for the Army Air Forces and the Navy.

3. Amson Rose Stimson Jr. was treasurer of the Dubois County Bar Association, secretary of the Huntingburg Airport, and chairman of the Aeronautics Committee of the Indiana Bar Association when he died in 1943 at the age of thirty-eight.

4. Janet Bednarek, "Fond of Flying: General Aviation," in *Taking Off: A Century of Manned Flight*, ed. Jonathan Coopersmith and Roger Launius (Reston, Va.: AIAA, 2003), 24.

5. David Cook, "Business Leaders Seek Major Military Air Base for Ocala," November 1, 2009, http://www.ocala.com/article/20091101/COLUMNISTS/911011013.

6. During World War II, George Clingman, later with the Aeronautics Commission of Indiana, bought, or conscripted, private aircraft to use for training military pilots. "Owners seldom parted with the aircraft amicably," according to his son David. Personal communication, 2013.

7. Air conditioning was uncommon at the time, and in mid-July 1943, people often slept outside because of the oppressive heat.

8. Clarence F. Cornish Interview, Douglas E. Clanin World War II Oral History Collection, 1944–2002, Collection M 0783, Manuscript and Visual Collections Department, William Henry Smith Memorial Library, Indiana Historical Society, Indianapolis. Clanin was an editor with the Indiana Historical Society. Between 1983 and 2002, he conducted interviews with World War II veterans. The Clanin collection includes both the original audiotapes of the interviews and later transcriptions; the transcripts of Cornish's interview, which was conducted on January 11, 1992, are in box 4, folders 11–12. The quotes from Cornish in the subsequent paragraphs are also from this interview.

9. *Mission Briefings* was a publication of the Heritage Museum Foundation, which founded the Grissom Air Museum in Peru, Indiana.

10. David Brinkley, *Washington Goes to War* (New York: Random House Value Publishing, 1988), 128.

11. For detailed information, Garry R. Pape with John M. and Donna Campbell, "Battlefield Florida" in *Northrup P-61 Black Widow: The Complete History and Combat Record* (Atglen, Pa.: Schiffer Publishing Ltd., 1995), 40.

12. Northrop ad, *Flying* magazine, July 1944, 123.

13. Unfortunately, when Mike later led a group of U.S. gliders into France, his craft slid across a dewy field, unable to slow down. He crashed into a tree stump and lay severely injured in a ditch for two days, "with the Germans parading up and down the road" next to him. In the movie *Saving Private Ryan* (1998), the depiction of the D-Day landing at Normandy Beach includes a fictionalized version of the crash.

14. H. Con. Res. 42, "Honoring the heroic service and sacrifice of the 6,500 glider pilots of the United States Army Air Forces during World War II," 100th Cong., 1st sess., January 23, 2007, http://www.govtrack.us/congress/bills/110/hconres42/text/ih. See also Michael H. Manion, "Gliders of World War II: 'The Bastards No One Wanted'" (master's thesis, Air University, School of Advanced Air and Space Studies, Maxwell AFB, 2008), http://www.dtic.mil/cgi-bin/GetTRDoc?AD=ADA493762.

15. During wartime, the U.S. military trained Chinese pilots at several locations, including Thunderbird Field in Glendale, Arizona.

16. "Some of the hazards which you may encounter include: unsafe conditions of landing fields or runways due to construction, repairs, snow, floods, etc.; installation or changes of lighting systems; failure in servicing facilities; congested training areas; aerial, bombing, and gunnery ranges; barrage balloon, searchlight, and certain construction areas; changes in operation of control towers and Army operated radio range or radio beam facilities." "Hazards to Air Navigation," in U.S. Army Air Forces, *Pilot's Information File* (Washington, D.C.: Army Air Forces, 1944–45), PIF 2-5-2.

17. *Instrument Flying Technique in Weather*, Technical Order no. 30-100D-1, Prepared by Authority of the Commanding General Army Air Forces, January 1, 1944. Available online at Army Air Forces Collection: Historical Documents from World War II, http://aafcollection.info/items/detail.php?key=148.

18. Four years later, returned to civilian life, Cornish established the National Association of State Aviation Officials's Washington office and recruited Colonel McMullen to be the executive director.

19. Author's collection.

Chapter 10

1. From Cornish's handwritten notes. Author's collection.

2. Clarence F. Cornish Interview, Douglas E. Clanin World War II Oral History Collection, 1944–2002, Collection M 0783, Manuscript and Visual Collections Department, William Henry Smith Memorial Library, Indiana Historical Society, Indianapolis. Clanin was an editor with the Indiana Historical Society. Between 1983 and 2002, he conducted interviews with World War II veterans. The Clanin collection includes both the original audiotapes of the interviews and later transcriptions; the transcripts of Cornish's interview, which was conducted on January 11, 1992, are in box 4, folders 11–12.

3. From Cornish's handwritten notes. In author's collection.

4. *Report on Facts, Findings and Recommendations*, Prepared by the Governor's Fact-Finding Committee on Aviation, C. F. Cornish, Chairman, June 22, 1939.

5. "Formation of State Air Board Asked," *News-Sentinel*, December 3, 1940.

6. Copy in author's collection.

7. Memo to Mr. Melvin Cohen from Postle dated August 18, 1944. Carbon copy in author's collection.

8. Cornish represented Indiana at NASAO's Western and Midwestern Airport Conference in Sacramento, California, in 1936. The attendees, who included representatives of military and civil aviation, the Bureau of Air Commerce, and aircraft manufacturers, worked on a plan to keep airport and aircraft development moving at an even pace. "Named Delegate," unidentified newspaper clipping. Cornish and Lois traveled to Sacramento as guests of Transcontinental and Western Air Inc.

9. American Association of Airport Executives, "History," http://www.aaae.org/about_aaae/history/.

10. Governor's Commission on Aviation, *Survey, Findings and Recommendations with Respect to Indiana's Aviation Problems* (Indianapolis: Governor's Commission on Aviation, 1944).

11. See "Acts of 1945, chapter 190, S.B. no. 114, A bill for an act authorizing municipalities to acquire, establish, construct, improve, equip, maintain and operate airports and landing fields . . . ," ibid., appendix 1.

12. Ibid., 1.

13. "Acts of 1945, chapter 360, H.B. no. 211, A Bill for an Act Creating an Aeronautics Commission of Indiana . . ." ibid., appendix 2.

14. "2 Indiana Air Groups Names: Will Represent Fliers on Policy Formation," *Indianapolis Star*, March 6, 1946, p. 3.

15. The various letters cited here are from the author's collection.

16. Personal communication.

17. Original document in author's collection.

18. Personal communication.

19. "Local Aviation and Airport Problems," talk presented by Cornish to the Indiana Commercial Secretaries' Association, November 25, 1946. Author's collection.

20. "Aeronautics Commission of Indiana, Fiscal Year 1949–50," in Indiana Division of Accounting and Statistics, *Year Book of the State of Indiana for the Year 1950* (Indianapolis: State of Indiana Executive Department, 1951), 748.

21. Ibid.

22. Merrill Swedlund, "Unique Stenographer Does Writing in Sky," *Spartanburg (S.C.) Herald*, December 28, 1949, 12; "Elizabeth 'Betty' Pettitt Nicholas," in *The Ninety-Nines: Yesterday—Today—Tomorrow* (Paducah, Ky.: Turner Publishing, 1996), 167.

23. The October 22, 1947, issue of *Air-Ads* included the following information

on "Markers for Airport Traffic Patterns Devised": "Landing an airplane at a strange field will be easier and safer with a new airport marker system which has recently been developed. The 'segmented circle' markers will be installed at all airports constructed under the Federal Airport Act and at all airports with right hand traffic patterns. Other airports will be urged to install the markers as an aid to visiting pilots. With the markers installed, pilots visiting a strange airport will be able to tell from the air the directions of the runways, and which runway is currently in use. Any unused traffic patterns at the airport, or the fact that the airport is closed will be evident at a glance. Best visibility was obtained by using a double panel erected like an inverted 'V.' Such markers are partially self-cleaning, can be seen for several miles laterally and tend to remain free of snow and mud." Author's collection.

24. *Indiana Aero-Notes* 1, no. 12 (October 1948).

25. Times State Service, Muncie, Ind., "Officials Applaud Air Defense Units' Work in 'Raid,'" July 16, 1951.

26. Municipal Airport Act of 1945, sec. IC 8-22-2-10.

27. *United States v. Causby*, 328 U.S. 256 (1946), Oyez Project, Chicago-Kent College of Law, http://www.oyez.org/cases/1940-1949/1945/1945_630.

28. Talk given by Cornish at the 4th Annual Aviation Conference, November 17, 1949, Severin Hotel, Indianapolis. Text in author's collection.

29. "FAA Historical Chronology, 1926–1996," faa.gov/about/media/b-chron.pdf.

30. Joyce's title is taken from a letter from Cornish dated December 12, 1952. In 1948 the CAA listed his official title and station as Nonscheduled Air-Carrier Inspector (Operations), Indianapolis, Ind., within the Region 3 Safety Regulation District Offices.

31. Helio Aircraft, "About Us: History," http://www.helioaircraft.com/aboutus_history.htm.

32. "Flying Farmers," *Air-Ads*, October 22, 1947, 8. *Air-Ads* was published by Herb Eichner of Royalton, Minn., "in the interests of the Central States Airman."

33. "On the Beam," talk given by Cornish to a Private Flyers Conference in Washington, D.C., November 26–27, 1945. Text in author's collection.

34. *Indiana Aero-Notes* 1, no. 7 (May 1948).

35. "Personal Flying in Indiana," speech given by Cornish at the Flying Farmer Clinic, Purdue University, December 30, 1948. Text in author's collection.

36. *Indiana Aero-Notes* 1, no. 1 (October 1947).

37. "On the Beam."

38. *Indiana Aero-Notes* 2, no. 3 (January 1949).

39. *Indiana Aero-Notes* 1, no. 2 (December 1947).

40. "Aeronautics Commission of Indiana, Fiscal Year 1945–46," in Indiana Division of Accounting and Statistics, *Year Book of the State of Indiana for the Year 1946* (Indianapolis: State of Indiana Executive Department, 1947), 214. A later document,

"The Federal-State Cooperative Aviation Safety Enforcement Program for the State of Indiana," founded upon the master CAA-CAB-NASAO Aviation Safety Enforcement Policy established in and dated February 1950, is in the author's collection.

41. Cornish talk at 4th Annual Aviation Conference.

42. Personal communication, ca. 2010. Cornish's files include mimeo copies of documents dated 1946 about early civilian use of ground controlled approach (GCA) radar landing systems, and "Approach System Evaluation Tests Conducted at Indianapolis." Author's collection.

43. On May 24, 1946, "The Civil Aeronautics Administration gave an initial demonstration of the first radar-equipped control tower for civilian flying atop the agency's Experimental Station at Indianapolis Municipal Airport. Raytheon had built the basic radar equipment for the Navy, and the company's engineers directed modifications at Indianapolis that included improvements lately developed for that service. Among these were an improved search antenna and a feature that eliminated ground clutter by permitting only moving targets to appear on the screen." "FAA Historical Chronology."

44. Copy of original proposal in author's collection.

45. Author's collection.

46. Copy of report in author's collection.

47. The Indiana Channeling Act of 1947 defined the methodology. Cornish explained the impact of the act on the Aeronautics Commission and its staff in a memo to Indiana senator John Van Ness dated March 1, 1949. Author's collection.

48. "Aeronautics Commission of Indiana," in Indiana Division of Accounting and Statistics, *Year Book of the State of Indiana for the Year 1948* (Indianapolis: State of Indiana Executive Department, 1949), 850.

49. Author's collection.

50. "Aeronautics Commission of Indiana," in Indiana Division of Accounting and Statistics, *Year Book of the State of Indiana for the Year 1947* (Indianapolis: State of Indiana Executive Department, 1948), 938. On July 30, 1947, President Harry Truman "signed Public Law 289, an amendment to the Surplus Property Act of 1944, to help speed the conversion to civil use of airports, airport facilities, and equipment no longer needed by the military. Recognizing that maintenance of the airports would require substantial funds, the law authorized transfer of surplus property to develop sources of revenue from non-aviation businesses at such airports." "FAA Historical Chronology."

51. "CAB Hearings Are Opened Here: Commercial Aviation Pattern Expected to Be Set," undated clipping from the *Indianapolis News*. It was agreed at the 1945 CAB meeting that a national air service system should be based on three equally essential parts: a primary pattern, to provide "inter-area scheduled air transportation"; a secondary pattern, "complementary to the Primary," to provide "scheduled inter-urban air service between communities whose relationship as to geographical location, population and commercial importance are such as to form natural routes for the logical, feasible and

active flow of air commerce between such communities"; and a tertiary pattern, "that part of the total air service system, not a part of either the primary or secondary systems, including but not limited to non-scheduled public air service."

52. Materials attached to Direct Testimony by Cornish before the CAB "In the matter of the great lakes area hearing of applications for proposed trunk and feeder line air service in the state of Indiana." Undated. Author's collection.

53. "The Role of the State in Air Transportation," speech given by Cornish at the Oklahoma University Feeder Airline Seminar, May 26, 1950. Text in author's collection.

54. Public Law 377, 79th Congress, 1946.

55. Original text of statement in author's collection.

56. "Aviation Officials Draft Report," *Indianapolis News*, February 5, 1948.

57. He was Indiana's governor from 1941 to 1945, but state law at the time did not allow governors to serve consecutive terms.

58. "Role of the State in Air Transportation."

59. Editorial, "Penny-Wise, Pound-Foolish," Michigan City *News-Dispatch*, February 6, 1951.

60. Karl Larsson, "Replacing the DC-3," *Flight Global*, November 13, 1953, 646.

61. Original copy of map in author's collection. Today there are four cities served by commercial airlines—South Bend, Fort Wayne, Indianapolis, and Evansville.

62. "What Goes On Here," *Indianapolis Times*, June 29, 1951.

63. Text of speech in author's collection. Founded as Roscoe Turner Airlines, Lake Central Airlines, with its home offices in Indianapolis, was a prominent carrier from 1950 to 1968.

64. "Makes Last Call with DC-3," *Indianapolis Star*, February 1, 1953.

65. Letter dated August 13, 1953. Author's collection. After devoting thirteen years of his life to the job, Cornish had been replaced by someone who would not remain at the helm for very long. After only a year, Krimendahl left the commission to join the sales team with Stokely-Van Camp, where his father was president and CEO.

66. Paul N. Janes, "Hoosier Air Travel Zooms Ahead," *Indianapolis Star Magazine*, August 23, 1953, 6–7.

67. Letter dated September 16, 1947. Author's collection.

68. As revealed in correspondence with Frank E. Kimble, Acting Chief, New Jersey Bureau of Aeronautics. Letter dated September 25, 1953. Author's collection.

69. Maurice Early, "The Day in Indiana," *Indianapolis Star*, early 1950s.

70. More than twenty-five years later, Cornish's contributions were still held in the highest regard. On May 6, 1979, he received a copy of *Art Smith's Story* as a gift from Bob McComb, a well-known aerobatics pilot. In it McComb had written: "With a long time admiration for his wide-angle, progressive views, planning and influence—on the civil, military and domestic scope—to help render the Aviation Industry to have become the great development as we know it today."

71. Letter dated October 27, 1953. Author's collection.
72. Letter dated December 7, 1953. Author's collection.
73. Author's collection.
74. Letter dated August 14, 1953. Author's collection.

Chapter 11

1. The quotations in this chapter are from personal communications.
2. The honorary Sagamore of the Wabash Award is the "highest honor which the Governor of Indiana bestows. It is a personal tribute usually given to those who have rendered a distinguished service to the state or to the governor. Among those who have received Sagamores have been astronauts, presidents, ambassadors, artists, musicians, politicians, and ordinary citizens who have contributed greatly to our Hoosier heritage." http://www.in.gov/core/files/Sagamore.pdf.
3. According to the International Forest of Friendship website at http://ifof.org/About_The_Forest.html, the memorial "was a gift to America on her 200th birthday in 1976 from the City of Atchison, Kansas (the birthplace of Amelia Earhart), The Ninety-Nines (International Organization of Women Pilots), and the University of Kansas Forestry Extension. . . . It is made up of trees from all fifty states and thirty-five countries around the world where Honorees reside."

Index

A

A. H. Grebe and Company, 48
Accidents, aviation, 74, 93–94, 194–195, 244n17
Adams, Noah, 211
Aereco, 68–75, 80–81, 82–83, 95
Aeronautics Commission Act, 182–184
Aero-Notes, 189, 194
Agricultural aviation, 192–193
Air-Age Education Program for the Elementary and Secondary Schools in Indiana, An, 188
Air circus events, 61–62, 98–100, 227n63
Air Commerce Act of 1926, 43–44, 91
Aircraft Manufacturer's Association, 58
Airmail, 42–44, 91–95
Air Mail Act of 1934, 95
Air marking, 195, 247–248n23
Airspace control and property rights, 190–191
Airview Addition, 136, 137
Aldous, William, 132
All-America Air Show, 119
Allen, James, 23

All Things Considered, 211
Amateur radio, 11–18, 24–25, 47–50, 60–61, 220n26
American Boy Scouts (ABS), 9–11
American Heritage History of Flight, 25–26, 31
American Legion air circus, 61–62, 227n63
American Newspaper Publishers' Association, 82
American Way, The, 134
Anders, Pete, 166
Aretz, Lawrence, 179
Armistice Day, 35
Army Air Force (AAF), 154, 156, 161, 242–243n6
Army Air Force in World War II, The, 151
Arnett, Richard, 109
Arnold, Henry H., 24, 166, 170, 172, 185
Art Smith Aeroplane Society, 218, 241n40
Ashley, Warner, 92
Auto Electric and Equipment Company, 77, 227n2

Auto Electric Radio Equipment
    Company, 65
Aviation. *See also* Smith-Baer Field
    accidents, 74, 93–94, 194–195,
        244n17
    accolades for Cornish in, 67–68
    Aereco and, 68–75, 80–81, 82–83
    agencies, 132–133
    agricultural, 192–193
    air circus events, 61–62, 98–100
    blind flying, 71, 228n8
    clubs and organizations, 58–59,
        73–74, 88, 97–101
    commercial sponsors of, 77, 90–91
    Cornish's early interest in, 8–9
    domestic air network and, 197–203,
        249–250n51
    education, 188–189
    end of Cornish's career in, 204–206
    experimental, 59–60
    famous achievements in, 67–68
    Federal Aviation Administration
        and, 92, 113, 130
    first solo flight of Clarence Cornish
        and, 1, 32–33
    Goral Airways and, 53–54, 56, 57, 68
    Immelman maneuver, 36–38
    Indiana Aeronautics Commission
        on, 179–186, 197–198, 206,
        207–208
    Indiana Aircraft Trades Association
        and, 97, 100–107
    mail transport via, 42–44, 91–95
    media coverage of, 52, 54–62, 73, 82
    military base conversions and,
        197–198
    military early interest in, 22–25
    mishaps of Clarence Cornish, 81
    navigational instruments, 81–82,
        173–174, 195–196, 235n69
    noise abatement and, 191–192
    police use of, 56
    popularity of, 97
    private, 193–194
    promotion by Clarence Cornish, 75,
        79–80, 188–189
    races, 90–91
    records set by pilots, 41–42
    safety in dangerous weather,
        173–174
    skills of Clarence Cornish, 34–38,
        51–52
    small airport development and,
        196–197, 201–202
    state legislation related to, 112–113
    training of Clarence Cornish, 26–32
    transatlantic, 76–77
    U.S. Signal Corps and, 21–26, 33
    women in, 67, 75, 78–79, 189, 232n1
Aviation Association of Indiana, 218
Ayers, Kenneth, 73

B

Baals, Harry, 108, 119, 123, 127, 166
    Smith Field and, 144–145
    U.S. Army Air Corps use of Baer
        Field and, 139–141
Babbitt, Amby, 98
Baer, Paul, 52–53, 225n26
Baer Field, 143–147, 165, 225n31
Baldwin, Irv, 151, 165–166
Ball, Edmund, xii, 82, 111, 235n61
Ball, Frank E., 109, 113
Balloons, barrage, 160–161
Baranyk, Steve, xv
Barksdale Field, 149–150, 156, 177
Barrage balloons, 160–161
Bartel, Robert R., 52, 115–116
Bass, John Henry, 5
Bass Foundry and Machine Company,
    17, 45
Bauer, George, 13, 16

Baur, Edith, 45
Baxter, Nancy Niblack, xv
Beamer, George M., 206
Bednarek, Janet, 167
Beeler, Jerome, 200
Bell, Laurence D., 205
Bell Aircraft Company, 205
Benninghof, Howard A., 117
Bern, Edward G., 119
Berry, J. H., 133
Biederwolf, Charles L., 138
Blind flying, 71, 228n8
Bollinger, Lynn, 192
Bombing, testing, and artillery ranges, 158–160
Borton, Mary, 80–81
Bostwick, Mary, 108, 112
Boucher, F. H., 140
Bowers, Charles, 113
Bowers, John W., 6
Bowling, Augusta "Lee," 176–177
Boyce, Walter, 67
Boy Scouts of America, 10, 219n12, 220n16
Brinkley, David, 87
Brock, William, 30–31, 67, 210
Brooks, Harold, 112
Brudi, Margaret L., 88
Buck, Charles W., 2–3
Buck, Louisa Durfee, 2, 3
Buck, William, 2–3
Burgoyne, Glenn, 81
Burgoyne Hardware Company, 81
Burnett, C. V., 132
Byrd, Richard, 67
Byroade, George L., 17

C

Campbell, Eugene W., 73, 90, 93
Campbell, Gene, 73
Canatsey, Andy, 73
Carlstrom, Victor, 35
Carter, George, 13–14, 47
Cassidy, Mid, 215
Cate, James Lea, 151, 154, 156
Causby, Thomas Lee, 190
Central Radio Association of Amateurs, 13
Cessna 170, 192
Cessna 172, x, 209, 211
Challenger Robin, 82
Chamberlin, Clarence D., 131
Chamberlin, Louise Ashby, 131
Champion Spark Plug Company, 77
Chandler, Charles, 23
Chicago Girls' Flying Club, 98
Chicago Tribune, 54
Citizens' Committee on Aviation, 139
Civil Aeronautics Act of 1938, 200
Civil Aeronautics Authority (CAA), 111, 130, 133, 137, 139, 140–141
   barrage balloons and, 161
   domestic air network design and, 197–199
   noise abatement and, 191–192
   state aeronautics commission and, 182
   World War II and, 158–159
Civil Aeronautics Board (CAB), 172, 186, 187, 197–198
Civil Air Patrol, 182
Civil Works Administration (CWA), 116
Clanin, Douglas, 56
Clark, Alex, 204
Clingman, George J., 188
Clubs and organizations, aviation, 58–59, 73–74, 88, 97–101, 132–133
Coats, Roscoe, 13
Collings, John, 121, 125
Commercial sponsors of aviation, 77, 90–91

Conner, Richard A. O., 73
Connolly, Donald H., 168
Consolidated Aircraft Repair Inc., 123
Contract Air Mail Act, 43–44
Cook, Weir, 73–74, 75, 111, 113
Copple, Mark, xv
Cornell, Beaumont, 184–185
Cornish, Ada Buck, 1, 2–7, 83
Cornish, Clarence Franklin. *See also* Military service of Clarence Cornish; Paul Baer Municipal Airport
    advocacy for private flying, 193–194
    Aereco and, 68–75, 80–81, 82–83
    amateur radio and, 11–16, 18, 47–50, 220n26
    American Boys Scouts and, 9–11
    American Legion air circus and, 61–62
    ancestors of, 2–3
    aviation mishaps of, 81
    birth of, 1
    childhood of, 7–8
    children of, ix, xi–xii, 131
    civil service record, 181
    early jobs held by, 11
    education of, xi–xii, 11
    end of career in aviation, 204–206
    experimental aviation and, 59–60
    first solo flight, 1, 32–33
    flying in later life, x–xi, 211–215
    flying skills of, 34–38, 51–52
    Goral Airways and, 53–54, 56, 68
    grandchildren of, xii, 212
    Indiana Aircraft Trades Association tour and, 97, 100–107
    in the insurance business, 209
    job searches after World War I, 44–45
    love of flying, 34–35
    as manager for Paul Baer Municipal Airport, x, 98, 115–116
    marriage of, 63–66
    media coverage of, 54–56
    military enlistment by, 19–20
    National Association of State Aviation Officials (NASAO) and, 199–200, 205, 208
    parents and siblings of, 2–7
    planes flown by, 94
    preserved correspondence and clippings of, xi, xv
    professional reputation of, 179
    promotion of aviation by, 75, 79–80, 188–189
    re-enlistment of, 149–150
    recognitions earned, 218
    Reserve Corps and, 50–52
    retirement of, 209–215
    social life of, 20–21, 45–46, 63–66
    state aeronautics commission and, 179–180
    teaching of other pilots by, 36
    Wayne Radio Company, 48–49, 65
    on women pilots, 78–79
    work in Washington, DC, 168–169
Cornish, Frank, 1, 3–7, 83
Cornish, Franklin, 2
Cornish, Irma Katherine, 3, 6–7, 45, 175
Cornish, John, 2
Cornish, Lois, 63–66, 77, 85–86, 108
Cornish, Marcia Odetta, 5–6
Cornish, Ruth Ann, 131, 211
Cornish, William, 2
Cornish & Bowers Company, 6
Cox, Charles E., 101, 107, 112
Cox, Paul, 109
Crabbs, Lucy Buck, 3
Craig, George, 204
Craven, Wesley Frank, 151, 154, 156

Crean, Tim, 211, 213
Crise, John W., 138
Crumpacker, Harry, 186
Cunningham, Richard L., xv, 186, 187–188, 195
Curtiss, Glenn, 8, 58
Curtiss JN-4D Jenny biplanes, 22, 26, 30–31, 211
    private purchases of U.S. Army, 41
Curtiss-Robertson Airplane Manufacturing Company, 75

D
Dale, W. Marshall, 139
Dawson, Gene, 182
Decker, Bill, 146
Decker, John F., 136
De la Cierva, Juan, 78
Detroit Aircraft Corporation, 88
Dicks, Herman, 56
Dienhart, Nish, 132, 180
Diffendorfer, Davis, 13
Doherty, Carl E., 89
Doolittle, Jimmy, 38, 75, 80, 82, 92, 228n8
Dorr, Stephen, 35
Douglas B-18M, 149–150
Droegemeyer, John L., 49–50
Duesler, B. J., 49
Dunman, M. W., 110

E
Earhart, Amelia, 67, 75, 78, 79, 232n1
Eastern Air Lines, 204
Education, aviation, 188–189
Egenroad, Charles L., 182
Ehrman, David G., xv
Eielson, Carl Ben, 75–76
Eisenhower, Dwight D., 204
Elliott, Richard Nash, 129

Elrod, Ed, xv
Erickson, E. H., 13
Evans, Floyd E., 132
Experimental aviation, 59–60

F
Fair, Ford, 53
Farley, James A., 94, 95, 119
Federal Aid Airport Program, 196
Federal Airport Act of 1946, 199
Federal Aviation Administration (FAA), 92, 113, 130
Federal Emergency Relief Administration, 107, 115, 118, 129
Federal Radio Commission, 48
Ferber, Ferdinand, 23
Fergerson, James, 73
Fisher, Herbert O., 75, 107, 108
Fleming, James R., 139
Flight attendants, 203–204
Flying Farmers, 192–193, 207
Flying Jenny. *See* Curtiss JN-4D Jenny biplanes
Flying the Lindbergh Trail, 99
Foellinger, Adolph, 86–87
Foellinger, Helene, 126, 210, 231n63
Foellinger, Loretta, 126
Foellinger, Oscar G., 59, 86–89, 90, 91, 230n48
Fokker plane, 57
Fort Benjamin Harrison, 51, 107, 114, 149
Fort Wayne, Indiana, 1. *See also* Paul Baer Municipal Airport
    airmail service and, 91–95
    air races in, 90–91
    aviation field, 52–53, 56–57, 65
    during the Great Depression, 84–85

Fort Wayne, Indiana (*continued*)
  Indiana Aircraft Trades Association tour and, 97, 100–107
  World War I and, 16–18
Fort Wayne Aero Club, 97–100, 132, 134
Fort Wayne Federation of Labor, 116
Fort Wayne Gas Company, 45
Fort Wayne Radio Association (FWRA), 13–16, 18
Foulois, Benjamin, 23
Francis Ferdinand, Archduke, 16
Frank, Eliza, 2
Frank Leslie's Popular Monthly, 7
Fulton, B. E., 132

G

Gallmeyer, Ruth, 45
Gates, Ralph F., 182, 185–186
Gatty, Harold, 76
General Electric, 16, 17, 138
General Motors, 53
Gerberding, Helen, 45
Geyer, Ben, 65
Gillie, George W., 139
Goral Airways, 53–54, 56, 57, 68
Goral Motor Company, 53
Gordon Bennett International Balloon Race, 119
Governor's Fact-Finding Committee on Aviation, 179–180
Grandt, Alten F., Jr., xii
Great Depression, 84–86, 129. *See also* Works Progress Administration (WPA)
Greeley, Phil, 166
Greenville Aviation, 168
Gregg, Whitney A., 107, 118
Grodian, Rachel F., 59
Grunert, Aro, 107
Guild, Paul, 92, 94–95, 107, 118, 119, 131

Guinness World Records, 218
Guinness Book of World Records, xi, 1, 212–213
Guy Means Airfield, 70, 72, 74, 77, 81–82, 87, 89, 92

H

Haizlip, Jimmy, 75, 92
Hall, Arthur F., 65, 84
Hamilton, Arthur G., 42
Hanley, Frank A., 168
Hartranft, Joseph B., 177
Hartzell, Lee, 63
Heckman, Mrs. E. W., 59
Helicopters, 205–206
Heliplane, 192
Helmke, Walter E., 118
Henninger, John, 73
Henry, Guy T., 182
Hermann, Louis, 13
Hill, George, 93
Hinton, Chester, 44
Hirschman, Alice, 108
Hitler, Adolf, 38, 128, 131, 152
Hobrock, Paul, 53
Hockemeyer, Bob, 175–176
Hockemeyer, Martin, 83
Holland Furnace Company, 45
Hollopeter, Herschel A., 180, 186
Hoover, Herbert, 86, 87
Hoover Airport, 87
Hosey, William J., 52, 57, 94
Hoyt, Jean Davis, 79

I

Illinois Air Pilots Association, 115
Immelman maneuver, 36–38
Indiana Aeronautics Commission, xii, xv, 179–186, 197–199, 206, 207–208
Indiana Aircraft Trades Association

(IATA), 97, 100–107, 179
Indiana Air Pilots Association (IAPA), 110–111
Indiana Aviation Association, 112
Indiana Chamber of Commerce, 182, 184
Indianapolis Aero Club, 189, 218
Indianapolis Chamber of Commerce, 107
Indianapolis Metropolitan Airport, x–xi
Indianapolis Municipal Airport, 107, 108
Indianapolis Star, 47, 108, 112, 205, 210
Instrument Flying Technique in Weather, 173–174
Instruments, navigational, 81–82, 173–174, 195–196, 235n69
Interdepartmental Air Traffic Control Board (IATCB), 150–151, 153, 155–156, 159, 173, 177
  barrage balloons and, 161
  labeling of restricted areas, 162–163
International Aircraft Show, 115
International Harvester Corporation, 84
International Organization of Women Pilots, 189
Interstate Commerce Commission, 125
Irmscher, Max, 45
Island Queen, 20

J

Jackson, Edwin T., 9, 75
Jernigan, W. P., 33
Johnson, Earl, 73
Johnson, Ed, 73
Johnson, Harold, 73
Jones, Matt, 67, 73
Journal-Chronicle, 60–61
Journal-Gazette, 11, 12, 14, 42, 46, 98, 108, 137
  on amateur radio, 49
  coverage of aviation, 52, 57
  on military enlistments, 17
  on the U.S. Army Air Corps and Paul Baer Municipal Airport, 138, 139
Joyce, Edwin A., 192

K

Kaffenberg, Harry, 122
Kavanaugh, Walter, 56
Keefer, Charles, 134–135
Kelly, Clyde, 43
Kelly Act, 43–44
Kerr, Marshall, 179
Kettering, Charles, 38
Kidd & Company, 6
Kirkpatrick, Clifford, 107
Klaehn, Robert, 89, 90
Knight, Jack, 43
Koch, Herbert V., 98
Koehlinger, Mildred, 45
Kokomo Airport, 107
Kramer's Business Directory, 2
Krieg, Christy, xvi, 212
Krieg, Lisa, xii, xvi
Krimendahl, David, 204
Kroger Grocery and Baking Company, 90

L

Ladies' Home Journal, 11
La Guardia, Fiorello H., 170
Lake Central Airlines, 203
Lassen, Warden, 89
League of Nations, 38
Legislation, state, 112–113
Lenz, Willard, 89
Leonard, George H., 132
Letter Shop, The, 6
Liggett, Harry, 211
Lindbergh, Ann, 88

260   Index

Lindbergh, Charles, 31, 48, 55–56, 59, 60, 67, 88
Littlejohn, John, 89
Lloyd, Bernard M., 73
Lohman, Eugene A., 143
Lombard, Carole, 244n17
Loomis, Ray, 107
Louis, Joe, 48
Louthan, Orville D., 73
Lovejoy, Frank, 53
Lusitania (ship), 16
Lyle, Jane, xvi

M

Macatee, W. R., 132
Mackin, Howard, 73
MacRobertson Air Race, 80
Mad Modelers, 132, 134
Magnavox, 84
Mahoney-Ryan Aircraft Corporation, 86
Mail transport, 42–44, 91–95
Malan, C. T., 188
Malone, George, 45
Manley, Don, xvi
Manufacturer's Aircraft Association (MAA), 58
Marriage of Clarence Cornish, 63–66
Marriott, J. S., 70, 174
Marshall, Jerry, 61
Mattern, Jimmie, 107
May, DeWitt, 13, 14–15
McComb, Bob, 80–81, 250n70
McCulloch, Ross, 108, 119, 125, 138
McElroy, Clarence, 110
McKittrick, Shirley, 108
McLaughlin, Joseph K., 208
McMullen, A. B., 150–151, 156, 174–175, 177
McVey, O. B., 77, 90, 92
Means, Guy, 53

Media coverage of aviation, 52, 54–61, 73, 82, 134–135. *See also Journal-Gazette; News-Sentinel*
Merrill, H. T., 204
Military base conversions, 197–198
Military service of Clarence Cornish
 aviation training during, 26–32
 end of World War I and, 38–39
 enlistment in, ix, 19–20
 first assignment, 20
 Morse code and, 28–29
 U.S. Signal Corps and, 21–26, 33
 World War II, 149–152, 156–157
Miller, Earl, 120, 127–128
Miller, Lanny, 211
Milling, Thomas, 24
Minton, Sherman, 139
Miser, Charles T., 201
Miss Champion, 77
Mission Briefings, 169
Mitchell, Billy, 35
Mitchell, Elsie, 154
Morgan, W. B., 122
Morse code, 28–29, 235n69
Moss, Earle W., 139
Muir, Ed, 187
Muncie Aviation Corporation, 109
Municipal Airport Act of 1945, 190
Murphy, Michael C., 75, 100, 107, 108, 109, 114, 172–173
Murphy, W. S., 60
Museum of Women Pilots, 78–79
Myers, Ben F., 92
Myers, Roger, xv, 93, 143, 146

N

National Aeronautic Association (NAA), 62, 76, 132–133, 188
National Aeronautics, 129
National Air Progress Week, 134

National Air Races, 90–91, 99–100, 109
National Air Week, 134
National Association of Airport Executives, 132
National Association of State Aviation Officials (NASAO), 199–200, 205, 208
National Public Radio, 211
National Radio Trade Association, 49
National Seminar on Feeder Airlines, 200
Navigation instruments, 81–82, 173–174, 195–196, 235n69
New Deal, 85
News-Dispatch, 201
News-Sentinel, 210
    on Cornish's management of Paul Baer Municipal Airport, 116
    coverage of air circus events, 100
    coverage of aviation industry, 45, 46, 56, 58–59, 61, 62, 73, 82, 83
    coverage of the All-Indiana Air Tour, 111
    on military air defense, 128
    on Paul Baer Municipal Airport, 123, 134–135, 139, 141–142
    on Wilkin's flight from Alaska to Norway, 76
    on World War II, 132, 133
    Yankee Clipper, 74, 86, 88–90, 91
New York, Chicago and St. Louis Railroad, 6
New York Central Railroad, 118
New York Times, 42
Nicholas, Betty, 211, 213–215, 217. *See also* Pettitt, Elizabeth
Nichols, Ruth, 78
Noise abatement, 191–192
Noyes, Blanche, 78
Nungesser, Charles, 52

O

O'Donnell, Gladys, 78
O'Donnell, Tim, 146
Offutt, Harry A., 73
Omlie, Phoebe, 78
Once upon a Canvas Sky, 25
Order of Daedalians, 218
Our Town Builds an Airport, 189–190
OX5 Aviation Pioneers, 218
    Hall of Fame for, 218

P

Pan-American Aeronautic Convention, 42
Parents and siblings of Clarence Cornish, 2–7
Parker, James E., 50
Parnin, Ross, 13, 16, 45, 46, 168
Pathfinder, 76, 77
Patterson-Fletcher Company, 73
Patterson-Fletcher Flying Club, 73–74
Patton, John, 52
Paul Baer Municipal Airport, x, 70, 73, 74–75, 76, 77, 80, 91, 95, 107, 110. *See also* Smith-Baer Field
    beer sold at, 117–118
    efforts to bring commercial airline traffic to, 119–121, 124–125
    housing development planned near, 135–136
    increase in traffic at, 119, 137–138
    public forum held at, 132–134
    TWA and, 118, 120–121, 125–128, 131
    two-way shortwave radio system, 120–121
    upgrades to, 91–92, 115–117, 122–128

Paul Baer Municipal Airport (*continued*)
    U.S. Army Air Corps and, 138–145
    Works Progress Administration and, 122–123
Paulsen, Margaret, 45
P-61 Black Widow, 171–172
Pénaud, Alphonse, 8
Pennsylvania Railroad Company, 14, 17
Perfection Biscuit Company, 11
Peru & Chicago Railway Company, 3
Peters, R. Earl, 95, 118
Pettitt, Elizabeth, 189, 192. *See also* Nicholas, Betty
Pierce, Lloyd, 90
Pitcairn, Harold F., 78
Pitcairn-Cierva Autogiro Company of America, 78
Pogue, L. Welch, 186, 188
Police use of aviation, 56
Porter, Robert, 9
Post, Wiley, 76, 229n22
Postal services, 42–44, 91–95
Postle, David A., 180, 201, 205
Potts, Clifford, 179
Pratt, John Emerson, 88
Preston, Harold, 54, 90, 92
Private aviation, 193–194
Property rights and airspace control, 190–191
Puff, Art, 45
Purdue Glider Club, 88
Purdue University, xii, 88, 182

Q
QST, 15, 220n26
Quinn, Charlie, 61

R
R. M. Kaough & Company, 77
Races, aviation, 90–91

Radio, 11–18, 24–25, 47–50, 60–61, 220n26
    Paul Baer Municipal Airport and, 120–121
Radio Act of 1927, 48
Rambling Recollections of Flying and Flyers, 82
Redifer, Rex, 25, 210
Reese, Jack, 192
Renshaw, Bill, 207
Reserve Corps, 50–52
Rickenbacker, Eddie V., 54, 75, 125
Rickenbacker Auto Company, 54
Ringle, Bernard J., 73
Rio Cruise, The, 134
Rock, Gene, 68, 107
Rockhill, Morrison A., 182
Roettger, Phillip, 200
Rogers, Will, 78, 177, 229n22
Rohr, Howard, 73
Romy, Fred, 167
Roosevelt, Franklin D., 85, 118, 121, 122, 128–129, 131–132, 154, 155, 177–178
Ross, George J., 204
Rust, Daniel, 118
Ryan, Oswald, 111
Ryan B-5, 76, 86, 88, 91, 94

S
S. F. Bowser and Company, 17
Sagamore of the Wabash Award, 211, 218, 251n2
Sarver, Paul, 45
Saturday Evening Post, 11
Schalliol, John, xvi
Scherer, Alma, 45
Scherer, Louis, 45
Schlee, Edward, 67
Schmitt, Alice, 45

Schmitz, Gilbert, 196
Schoen Field, 107, 114
Schott, Bob, 73
Schott, Robert T., 73
Schricker, Henry F., 180, 200
Schroeder, R. W., 41–42
Scott, Blanche Stuart, 8
Shaut, Earl W., 211
Shirra, Walter, Jr., 215
Shoemaker, Carolyn, 215
Shoemaker, Eugene, 215
Siebert, Kenneth, 53
Slagle, Verne, 48
Small airport development, 196–197, 201–202
Smith, Art, 8, 119, 132, 145
Smith, C. R., 125
Smith, Stanton T., 107
Smith-Baer Field, xvi, 143–147, 165–166
    conflicts at, 165–167
    pilot training at, 167
Smithsonian Book of Flight, 67
Social life of Clarence Cornish, 20–21, 45–46, 63–66
Spurgeon, K. A., 168
Starr, George W., 182, 186
Stevens, A. W., 42
Stewardesses, 203–204
Stimson, A. R., 167–168
Stratemeyer, George E., 161–162
Straub, Elmer, 180
Strobel, Carl, 45
Sturm, W. F., 101
Sweebrock Airport, 54
Sweet, Bill, Jr., 100, 109
Sweet, George, 53

T
Tarkington, Alvin, 110

Telegraphy, wireless, 11–16, 18, 24–25, 47–50
Thaden, Louise, 78
Thomas-Morse Scout, 36, 37
Todd, Levi, 63
Townsend, Clifford, x, 179, 180
Transatlantic aviation, 76–77
Travel Air Cabin, 82
Truman, Harry, 177–178
Turner, Roscoe, 75, 100, 131, 185, 215
TWA, 74, 118, 120–121, 125–128, 131, 165, 244n17

U
United Airlines, 120
U.S. Army Air Corps, x, 93–94, 98, 138–145, 242–243n6
U.S. Army School of Military Aeronautics, 27
U.S. Department of Commerce, 70–71
U.S. Postal Service, 42–44
U.S. Signal Corps, 21–26, 33
U.S. Weather Bureau, 133, 135, 155

V
Van Nuys, Frederick, 138
Van Orman, Ward T., 120
Voetler, Karl E., 206
Von Braun, Werner, 38

W
Wall, Herbert C., 65
Walters, Ralph, 73
Ward, Clifford B., 119
Washington Airport, 87, 168
Washington Evening Star, 23
Waterloo Press, 75
Watson, Chester K., 113

Watson, Jack, 170
Watterson, Florence, 64
Watterson, Lois Lucile, 63–66. *See also* Cornish, Lois
Wayne, Anthony, 145
Wayne Radio Company, 48–49, 65
Webster, Grove, 141
Weir Cook Airport, 192, 200
Wells, Bill, 209
Wells, Harold, 209
Wells Insurance Agency, 209
Welsheimer, William A., xvi
White, Don, 98
White, Jane, 2
Who's Who in Transportation and Communication, 218
Wildwood Builders Company, 6
Wilkins, Hubert, 75–76
Williams, Morton B., 13
Williams, Roger A., 76
Williams, Ted, 170
Wilson, Woodrow, 16, 22
Winnie Mae, 76
Winslow, Walker, 73, 180
Wolf, Martin J., 73
Women in aviation, 67, 75, 78–79, 189, 232n1
Wonder Book of Aircraft for Boys and Girls, The, 30
Works Progress Administration (WPA), 118, 122–123, 138
World War I, ix, 16–18. *See also* Military service of Clarence Cornish
    end of, 35, 38
    growing use of aviation in, 22–24
World War II, 38, 132, 133
    barrage balloons used during, 160–161
    bombing, testing, and artillery ranges during, 158–160
    Cornish's summary of his work during, 177
    Cornish's work in Washington, DC during, 168–169
    homeland defense during, 154–158, 170–171
    Pearl Harbor attack and, 152, 160, 173
    U.S. lack of preparation for, 152–155
Wrib, Ed, 166
Wright, Milton, 8
Wright, Orville, 8, 22, 23, 58
Wright, Roderick M., 204
Wright, Wilbur, 8, 22
Wright Flyer, 8, 23
Wustenfield, Richard W., 65

Y

Yancey, Lewis A., 76–77
Yankee Clipper, 74, 86, 88–90, 91, 118
Young, L. C., 13
Young, Russell, 86

www.ingramcontent.com/pod-product-compliance
Lightning Source LLC
Chambersburg PA
CBHW071701160426
43195CB00012B/1538